50 Kubernetes Concepts Every DevOps Engineer Should Know

Your go-to guide for making production-level decisions on how and why to implement Kubernetes

Michael Levan

‹packt›

BIRMINGHAM—MUMBAI

50 Kubernetes Concepts Every DevOps Engineer Should Know

Group Product Manager: Rahul Nair

Publishing Product Manager: Niranjan Naikwadi

Senior Editor: Tanya D'cruz

Technical Editor: Rajat Sharma

Copy Editor: Safis Editing

Project Coordinator: Ashwin Kharwa

Proofreader: Safis Editing

Indexer: Rekha Nair

Production Designer: Nilesh Mohite

Senior Marketing Coordinator: Nimisha Dua

First published: February 2023

Production reference: 1130123

Published by Packt Publishing Ltd.

Livery Place

35 Livery Street

Birmingham

B3 2PB, UK.

ISBN 978-1-80461-147-0

www.packt.com

To my son, Zachary, for one day, once you're older, understanding why I work so hard, and to my mother for always being there to help out. To the community – thank you for enjoying the work that I put out and taking this journey with me.

– Michael Levan

Contributors

About the author

Michael Levan is a seasoned engineer and consultant in the Kubernetes space who spends his time working with start-ups and enterprises around the globe on Kubernetes and cloud-native projects. He also performs technical research, creates real-world, project-focused content, and coaches engineers on how to cognitively embark on their engineering journey. He is a DevOps pro, a HashiCorp Ambassador, and an AWS Community Builder, and loves helping the tech community by speaking internationally, blogging, and authoring technical books.

About the reviewer

Chad Crowell has been in the tech industry for 15 years, working as an Engineer, a DevOps Consultant, Kubernetes Instructor, and a Microsoft Certified Trainer. Chad has also authored the book *Acing the Certified Kubernetes Administrator Exam*. He is passionate about helping others overcome obstacles in their life and work, and embraces the community and open-source aspects of working in teams.

Table of Contents

Part 1: First 20 Kubernetes Concepts – In and Out of the Cloud

1

Part 2: Next 15 Kubernetes Concepts – Application Strategy and Deployments

5

Deploying Kubernetes Apps Like a True Cloud Native   103

6

Kubernetes Deployment– Same Game, Next Level   135

Part 3: Final 15 Kubernetes Concepts – Security and Monitoring

7

Kubernetes Monitoring and Observability   171

8

Security Reality Check   207

Preface

The idea behind Kubernetes is to make engineers' lives easier, right? Although true, there are pros and cons to every technology and platform. At the end of the day, Kubernetes does make handling containerization more efficient, but that doesn't mean that it's easy. Many organizations and engineers put in a lot of effort to truly get Kubernetes running the way it should run.

The goal of this book, and the overall 50 concepts, is to help mitigate some of these headaches. Although one book cannot mitigate every single issue that can occur, or make every single component work the way that it's supposed to, the overall goal is to help you use Kubernetes in an easier fashion in production, with 50 key pieces ranging from cloud to on-prem to monitoring and security, and everything in between. The world is currently full of content and ways to teach you Kubernetes. This book is to help you make it to the next level.

Throughout this book, you'll see everything from creating environments to deploying a service mesh and Kubernetes resources. I won't lie – a lot of the topics in this book are literally books in themselves. Because of that, the explanations and overall pieces had to be trimmed down a bit. Because of that, you may not have all of the answers in this book, but it'll give you an extremely good place to start your Kubernetes production journey.

With the 50 concepts in mind, you should be able to take what you learn here and ultimately expand on it in your production environment. Take what you learn, apply it, and ultimately, know which direction to go in to learn more about the concepts.

Who this book is for

This book is for the engineer that wants to use Kubernetes in production. Perhaps you've just learned the basics and beginner-level information about Kubernetes, and you're now ready to make it to the next level. Maybe you're getting ready to implement Kubernetes in production or test out containerized workloads for your environment. In either case, you can use this book as a source to showcase what you should be thinking about in production.

Think about this book as almost a "guide." There's theory, hands-on sections, and actual code that works from start to finish to create and deploy Kubernetes resources. As mentioned in the preface, this book can't cover every single topic in depth, as many of the topics are books within themselves, but you can use it as a "guide" to deploy to production.

What this book covers

Chapter 1, *Kubernetes in Today's World*, goes over, from a theoretical perspective, how you should think about Kubernetes in the current ecosystem – things such as why it's important, what the whole idea of "cloud native" means, and what containerization as a whole is doing for engineers.

Chapter 2, *Getting the Ball Rolling with Kubernetes and the Top Three Cloud Platforms*, hits the ground running with cluster deployments. You'll learn how to deploy Kubernetes clusters in Azure, AWS, and GCP. You'll see from a UI/GUI perspective how to deploy the clusters with code. This chapter uses Terraform for **Infrastructure as Code (IaC)**, as that's currently the most popular method in production.

Chapter 3, *Running Kubernetes with Other Cloud Pals*, teaches you how to deploy the top three most popular managed Kubernetes services. However, that doesn't mean those are the only methods. In this chapter, you'll see a few more popular options that are used in production but are mostly used for testing production workloads, as they're a bit cheaper from a cost perspective.

In today's cloud-centric world, a lot of technical marketing and content that you see on social media doesn't talk about on-prem. The reality is that on-prem, especially on-prem Kubernetes clusters, are still very much a thing. In *Chapter 4*, *The On-Prem Kubernetes Reality Check*, you'll learn about how to think about on-prem from a theoretical perspective and a bit hands-on.

Chapter 5, *Deploying Kubernetes Apps Like a True Cloud Native*, starts your journey into deploying applications to the cloud. In the first few chapters, you learned about cluster management, which is drastically important but only one half of the puzzle. The second piece of the puzzle is actual Kubernetes resource deployment.

Starting off where you left off in the previous chapter, *Chapter 6*, *Kubernetes Deployment – Same Game, Next Level*, takes Kubernetes resource deployments to the next level. You'll be introduced to concepts such as CI/CD, GitOps, and service mesh deployments. This is considered the "advanced" piece of Kubernetes resource deployments, which you'll see a lot of in production.

Up until this point in the book, you've learned how to deploy and manage clusters and applications. Once clusters and apps are deployed, you then need to confirm that they're running as expected. That's where observability and monitoring come into play, which we will look at in *Chapter 7*, *Kubernetes Monitoring and Observability*.

To wrap up any Kubernetes production deployment, you need to think about one major element before any resource reaches production – security. Security is the make or break between a successful environment and a long weekend of putting out fires. In *Chapter 8*, *Security Reality Check*, you'll learn the major components to secure a Kubernetes environment and a few key tools and platforms that you can use to make it happen.

To get the most out of this book

This book is a healthy combination of theory and hands-on. The reason for this is that theory is great, but if you don't know how to implement it, it's not going to be much use to you in production. To follow along with this book, you should have access to the major clouds, a few VMs, and perhaps a few dollars to spend on the environments.

Software/hardware covered in the book	Operating system requirements
Kubernetes v1.24 and above	Windows, macOS, or Linux

If you are using the digital version of this book, we advise you to type the code yourself or access the code from the book's GitHub repository (a link is available in the next section). Doing so will help you avoid any potential errors related to the copying and pasting of code.

Download the example code files

You can download the example code files for this book from GitHub at `https://github.com/PacktPublishing/50-Kubernetes-Concepts-Every-DevOps-Engineer-Should-Know`. If there's an update to the code, it will be updated in the GitHub repository.

We also have other code bundles from our rich catalog of books and videos available at `https://github.com/PacktPublishing/`. Check them out!

Download the color images

We also provide a PDF file that has color images of the screenshots and diagrams used in this book. You can download it here: `https://packt.link/FQMAS`.

Conventions used

There are a number of text conventions used throughout this book.

`Code in text`: Indicates code words in text, database table names, folder names, filenames, file extensions, pathnames, dummy URLs, user input, and Twitter handles. Here is an example: "Mount the downloaded `WebStorm-10*.dmg` disk image file as another disk in your system."

A block of code is set as follows:

```
terraform {
  required_providers {
    azurerm = {
      source  = "hashicorp/azurerm"
    }
```

```
    }
}
```

When we wish to draw your attention to a particular part of a code block, the relevant lines or items are set in bold:

```
variable "name" {
    type = string
    default = "aksenvironment01"
}
```

Any command-line input or output is written as follows:

```
sudo systemctl daemon-reload
sudo systemctl enable crio --now
```

Bold: Indicates a new term, an important word, or words that you see on screen. For instance, words in menus or dialog boxes appear in **bold**. Here is an example: "Select **System info** from the **Administration** panel."

> **Tips or important notes**
> Appear like this.

Get in touch

Feedback from our readers is always welcome.

General feedback: If you have questions about any aspect of this book, email us at customercare@packtpub.com and mention the book title in the subject of your message.

Errata: Although we have taken every care to ensure the accuracy of our content, mistakes do happen. If you have found a mistake in this book, we would be grateful if you would report this to us. Please visit www.packtpub.com/support/errata and fill in the form.

Piracy: If you come across any illegal copies of our works in any form on the internet, we would be grateful if you would provide us with the location address or website name. Please contact us at copyright@packt.com with a link to the material.

If you are interested in becoming an author: If there is a topic that you have expertise in and you are interested in either writing or contributing to a book, please visit authors.packtpub.com.

Share Your Thoughts

Once you've read *50 Kubernetes Concepts Every DevOps Engineer Should Know*, we'd love to hear your thoughts! Scan the QR code below to go straight to the Amazon review page for this book and share your feedback.

https://packt.link/r/1804611476

Your review is important to us and the tech community and will help us make sure we're delivering excellent quality content.

Download a free PDF copy of this book

Thanks for purchasing this book!

Do you like to read on the go but are unable to carry your print books everywhere?

Is your eBook purchase not compatible with the device of your choice?

Don't worry, now with every Packt book you get a DRM-free PDF version of that book at no cost.

Read anywhere, any place, on any device. Search, copy, and paste code from your favorite technical books directly into your application.

The perks don't stop there, you can get exclusive access to discounts, newsletters, and great free content in your inbox daily

Follow these simple steps to get the benefits:

1. Scan the QR code or visit the link below

https://packt.link/free-ebook/9781804611470

2. Submit your proof of purchase
3. That's it! We'll send your free PDF and other benefits to your email directly

Part 1:
First 20 Kubernetes Concepts –
In and Out of the Cloud

When engineers first dive into Kubernetes, it can almost feel like a tool of sorts. You use it to run and deploy containers. However, that's not the case. Kubernetes in itself is a platform. It's almost like a subset of a methodology to run containers. Kubernetes, among many other platforms, is why the whole *platform engineer* title is becoming so popular. The DevOps space is moving away from thinking about just tools and focusing on the entire platform and environment.

With different platforms comes the question – where do you run it? The first set of concepts in the 50 concepts will be explained here with the overall architecture of Kubernetes.

Kubernetes is something that's heavily utilized in the cloud, but it's also heavily utilized on-premises. An example of this is certain regulatory requirements. I was recently chatting with a colleague that works in the defense space. Because of the obvious heavy security requirements that they have, along with the need to stay as close to certain areas as possible, using Kubernetes at the edge was a crucial part of success. Some of the edge Kubernetes nodes were running on k3s, which is a popular method of running Kubernetes for ARM devices. Those ARM devices are on-premises, so not in the cloud.

On the flip side, a lot of organizations don't have this regulatory requirement, so running Kubernetes in the cloud as a managed service is perfectly valid. It's also an easier approach to hit the ground running with Kubernetes. For example, utilizing **Azure Kubernetes Service (AKS)** is a lot easier from the start than bootstrapping a five-node cluster with Kubeadm.

By the end of this part, you should fully understand how to get started with Kubernetes, how to run it in the cloud, and the different popular cloud services that are available for you to use. Although not every single managed Kubernetes service is covered in these four chapters, these chapters will get you through the most popular along with giving you a solid idea of how the other services will look and be utilized.

This part of the book comprises the following chapters:

- *Chapter 1, Kubernetes in Today's World*
- *Chapter 2, Getting the Ball Rolling with Kubernetes and the Top Three Cloud Platforms*
- *Chapter 3, Running Kubernetes with Other Cloud Pals*
- *Chapter 4, The On-Prem Kubernetes Reality Check*

Kubernetes in Today's World

If you're reading this book, chances are you've been, or still are, in the tech/IT world in some capacity, whether it's from the operations side, the development side, or both – perhaps even technical leadership or product management. In any case, you've most likely heard about a platform/technology called **Kubernetes**. From how every company, both small and large, is talking about Kubernetes, a lot of engineers and leadership personnel think it's going to solve many problems. Although that's true, there's a twist, and with everything that makes our lives easier, there are caveats.

This chapter is primarily theoretical and will answer a lot of the questions you most likely have about moving to the cloud, hybrid environments, cloud-native/specific applications, and how Kubernetes is taking over the microservice ecosystem.

By the end of this chapter, you'll be able to answer some of the questions about the pros and cons of implementing Kubernetes. You'll have a solid understanding of why engineers and leadership teams alike are moving to Kubernetes. The gears will also start moving in your head concerning what your current application(s) look like and whether Kubernetes would be a good fit.

In this chapter, we're going to cover the following topics:

- The shift to the cloud
- Kubernetes, the new cloud OS and data center
- Cloud-native applications and why they're important
- Abstraction is easier, but with a twist
- Start slow and go fast

Technical requirements

This chapter will be more theory than hands-on, so you don't have to worry about any coding or labs. To follow along with this chapter, and this book in general, you should have beginner-level knowledge of Kubernetes, intermediate knowledge of the cloud, and some experience with applications and architecture.

Moving forward, a lot of the chapters in this book will include labs and hands-on work. You can find the code for each exercise in this book's GitHub repository at `https://github.com/PacktPublishing/50-Kubernetes-Concepts-Every-DevOps-Engineer-Should-Know`.

The shift to the cloud

Before diving into Kubernetes, there's an important question to ask: Why use Kubernetes? The reality is that organizations have been deploying applications without Kubernetes for years. There wasn't Kubernetes in the 1980s when engineers were writing software to floppy disks. So, why now?

The answer to this question is a complicated one and the best place to start is by thinking about what the cloud does for us – not necessarily what the cloud is, but instead what the cloud helps us think about when deploying software and systems from an architect, engineering, and management perspective. In this section, you're going to learn about the following aspects of the cloud:

- Why organizations care about the cloud
- What the cloud did for engineers
- How abstraction can help us learn from our mistakes
- How the cloud doesn't exactly do what people think it does

Let's take a closer look.

Why organizations care about the cloud

Leadership teams in organizations, whether it's the CIO, CTO, or someone in a technical leadership position, tend to tie Kubernetes to the cloud. However, this couldn't be any further from the truth. The reason why could be anything from incredibly good technical marketing to not having enough experience from a hands-on perspective to truly understand what's happening underneath the hood in a cloud environment. However, let's digress from that and think about why everyone cares about the cloud. The best way to do this is with a visual, so let's take a look at the following diagram:

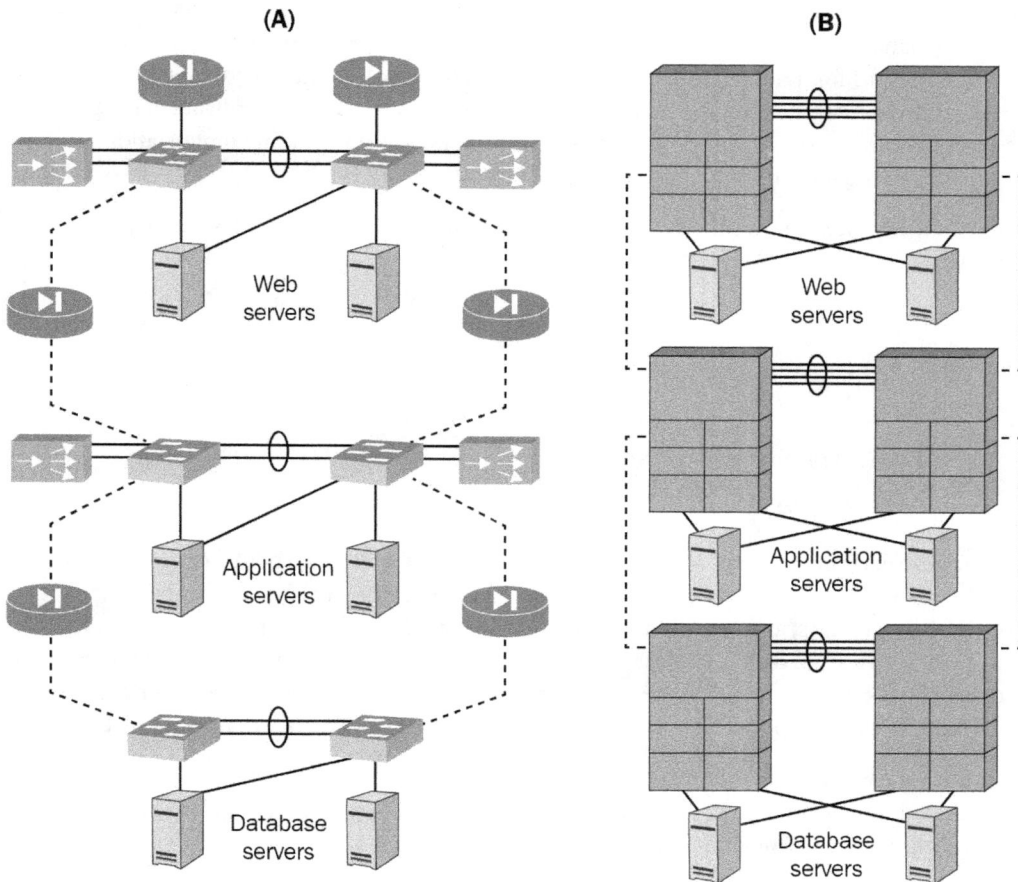

Figure 1.1 – Data center web app architecture

The preceding diagram is of a data center architecture. There are a lot of pieces, some marked and some not, including the following:

- Servers
- Network equipment
- Connections between servers
- Ethernet cables and power cables

With all of that hardware not only comes costs of actually buying it, but also costs around hiring engineers to manage it, maintain it, and keep the lights on in the data center. Not to mention it takes about 4 to 6 months for a full data center to be up and running. With the time that it takes the data center to get up and running, on top of all of the costs and management of hardware, having cloud-based systems starts to make a lot of sense to senior leadership teams for any organization ranging from start-ups to the Fortune 500.

Now, let's take a look at the architecture in *Figure 1.2*. This diagram shows a few things, including the following:

- RDS (Amazon's database service)
- Load balancers
- CDNs
- S3 buckets (cloud storage in AWS)
- Route 53 (AWS DNS)

The architecture diagram in *Figure 1.2* is similar to *Figure 1.1*, in that they are both data centers, but only *Figure 1.2* is virtualized as exactly what you would see in a data center. You have network components, storage, databases, servers, and everything in between. The biggest difference is what you're seeing here is virtualized. It's a virtual data center with virtual services. Because there are engineers that work at AWS managing the hardware, networking, and other peripherals for you, you no longer have to do it. You only have to worry about the services themselves and making sure they're working as expected.

No more buying hardware. No more replacing hard drives. No more waiting 4 to 8 months for hardware to arrive at your data center so you can finally build it. Instead, an entire data center is only a few button clicks or a few lines of automation code away:

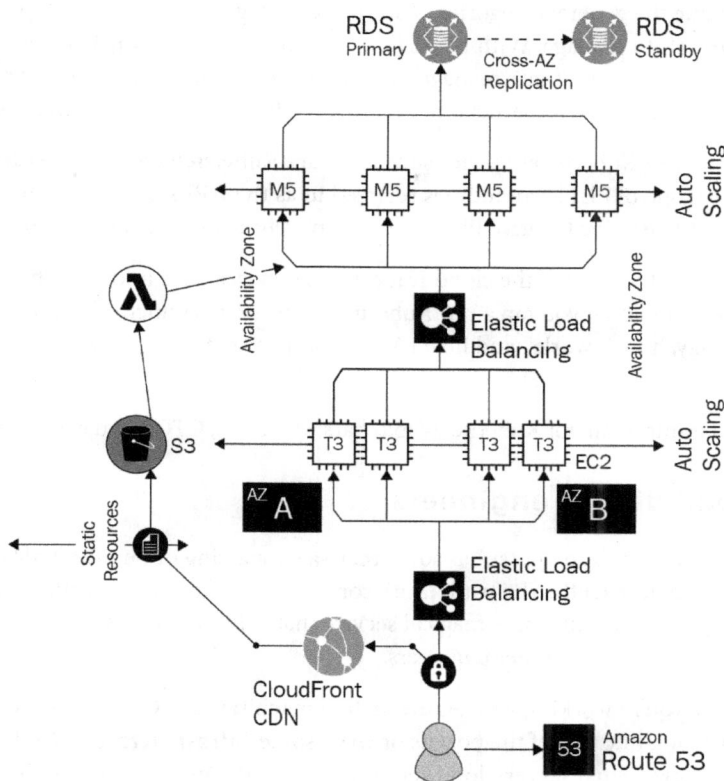

Figure 1.2 – Cloud web app architecture

Going off of the preceding diagram, here's where Kubernetes comes into play. Regardless of what option you go with, on-premises or in the cloud, there's still a ton of stuff to manage. Even though the cloud makes infrastructure easier, there are still major staffing needs and a big cost behind creating all of the cloud infrastructures.

The following are a few pieces of the puzzle to manage:

- Load balancers
- Virtual machines (or bare-metal servers)
- Network equipment (virtual or physical)
- Subnets, private IPs, public IPs, and gateways
- Security for multiple pieces of virtualized hardware and services

And that's just to name the general categories. Within each category, there are multiple components (physical and/or virtual) to manage. With Kubernetes, it's all abstracted away from you. There aren't any firewalls or gateways because that's managed via kube-proxy. There are no virtual machines that you have to deploy other than the Kubernetes nodes because the apps are running in Kubernetes Pods.

If you run Kubernetes in a Kubernetes service such as **Azure Kubernetes Service (AKS)** or GKE, the management of the Control Plane, sometimes referred to as the API server or the master nodes (a now deprecated way to describe Control Planes), is completely abstracted away from you.

What AKS, GKE, or another one of the cloud Kubernetes services does underneath the hood is the same thing that you would do if you ran a raw Kubernetes cluster in a bunch of virtual machines. The underlying technology, how it works, and how it's used don't change. The only thing that changes is the abstraction.

That's why the cloud is important for Kubernetes and that's why CIOs, CTOs, and engineers should care.

What the cloud did for engineers

"The cloud is just someone else's computer," as some may say in passing or on funny stickers. As we all know, in every joke is a little truth. The truth is, it's correct. When you're interacting with the cloud, it's not that you're interacting with some magical service that is *just there*. Instead, you're interacting with services that are managed by other engineers.

For example, let's say you're working with Azure virtual machines or EC2 instances in AWS. From your perspective, you log into one of the portals or write some **Infrastructure as Code (IaC)** and in a few minutes, your new virtual server/cloud server is deployed. On the backend, there's way more that goes into it. There are a ton of different parts, some of which include the following:

- Autoscaling the servers
- Doing multiple checks to ensure that there's enough hard disk, CPU, and RAM space on the physical/bare-metal server that's being used
- Networking setup
- Lots of automation

Remember, because the cloud servers that you're deploying are running on bare-metal servers, people have to be there to manage those servers and maintain them. The cloud is an abstraction layer that you don't see. With that being said, the cloud has done a lot for engineers.

Let's take a start-up company for example. Years ago, if a start-up company wanted to do anything in the tech space, they needed servers to host applications and websites. For a small company that's working out of a tiny office or even from someone's house, it's not possible to have a layer of high availability, redundancy, and scalability. They simply cannot afford the hardware, the space, and the employees to do it.

With the cloud, they no longer have to worry about having to do all of that. Instead, the start-up can focus on building applications and deploying them to the cloud. Deploying applications to the cloud is not easy and it certainly has its own complexity, but the idea behind it is to abstract away physical needs (servers, infrastructure, and so on) that your company may not want to/have to worry about.

Kubernetes, the new cloud OS and data center

Kubernetes is a topic that's on everyone's mind, but at the same time, a lot of individuals don't understand why. Is it the actual platform itself? Or what the platform does for engineers in today's world? The answer to those questions is – sort of both. Kubernetes does several things, but the primary pieces include the following:

- Deploying your containerized application

- Scaling your application

- Ensuring that your application is highly available

- Giving you the ability to secure your application and the users accessing the application

These four points sound like what engineers have already been doing with computers since the inception of the first mainframe. The question now becomes, why is Kubernetes so popular?

Kubernetes in the cloud

Everywhere you look, it feels like there's a new way to utilize the Kubernetes platform or some new tool that's supposed to make your life easier. Some of these platforms include the following (you'll learn more about these in the upcoming chapters):

- Cloud Kubernetes services such as AKS, **Google Kubernetes Engine (GKE)**, and Amazon **Elastic Kubernetes Service (EKS)**

- **Platform-as-a-Service (PaaS)** offerings such as OpenShift

- Serverless Kubernetes platforms such as Azure Container Apps and AWS Fargate profiles on EKS

Although that's not an extensive list, you can see that just the sheer number of platforms that are at your disposal can make it extremely difficult to pick and choose what you should ultimately go with. The semi-agreed-upon answer to this question is that it all depends on your current ecosystem. If you're in AWS, use EKS. If you're in Azure, use AKS. If you're a Red Hat Enterprise customer, check out OpenShift. The reason why is that, at the end of the day, all of the Kubernetes services are doing the same thing. They're all using Kubernetes under the hood and utilizing cloud services to make your life easier.

For example, if you're using AKS, chances are you probably want to use **Azure Active Directory (AAD)** to manage who has access to what in the AKS cluster. Azure makes it extremely straightforward to implement this because the goal of a Kubernetes service in the cloud is to do exactly that. All public clouds in general are trying to make your life easier, regardless of what cloud you're using. A great example of this is how you can use AAD inside of GKE via federation with Cloud Identity to map AAD tenants, users, and groups.

Why Kubernetes?

The question from the beginning of this chapter around why people want to use Kubernetes has sort of been answered, but there's still more to think about. Primarily, we must think about why everyone is flocking toward Kubernetes, especially Kubernetes services in the cloud. The answer to why people are using Kubernetes services in the cloud is typically something similar to one of the following:

- You don't have to worry about the underlying infrastructure
- Worker nodes and Control Planes are scaled for you *automagically*

And although those are great answers, you're still not any closer to the answer as to why you should use Kubernetes if all it's doing is what everyone has been doing in tech for years. It's not implementing anything new or out of the ordinary.

Simply put, the reason why people like Kubernetes is that it allows you to interact with your infrastructure via an API. When you run a Kubernetes command such as `kubectl apply -f deployment.yaml`, you're interacting with the Kubernetes API. When you run a command such as `kubectl get deployments`, you're interacting with an API. 99% of what you do when interacting with Kubernetes is all API-based. It's a bunch of `GET` and `POST` requests. The reason why Kubernetes makes engineers' lives easier is that what you used to have to do to get an application up and running on multiple servers is now abstracted away and it's all now at the programmatic level. All APIs.

Kubernetes as a data center

Remember data centers? Those things that have the loud, big computers running with a bunch of fans and air conditioners? Perhaps you're from the era of spending hours in a data center, racking and stacking servers, and taking a nap on the data center floor using your backpack as a pillow. If you've never done any of that, consider yourself a lucky person!

When thinking about a data center, there are several components, but let's think about the main ones that engineers care about:

- Servers
- Network equipment (firewalls, load balancers, routers, switches, gateways, and so on)

- Outbound and inbound connectivity
- Security
- The ability to run software and virtualization on the servers

Containerization platforms such as LXC and Docker were able to give us the fifth point mentioned here – virtualization of OSes and the ability to run software – but what about the rest? Engineers needed a way to orchestrate and manage the software and virtualized OSes. That's where Kubernetes comes into play.

Kubernetes fills every piece of the data center puzzle:

- Networking, including Pod-to-Pod communication, services, service meshes, Ingress, load balancing, and routing.
- Security and encryption between Pods and services
- High availability for clusters
- The ability to deploy, manage, scale, and maintain applications of any kind (must be containerized)
- Authentication and authorization capabilities from third-party tools such as AAD and IAM users/roles

Kubernetes is a one-stop shop for everything that you would find in a data center. The biggest difference is that the infrastructure (if you're running in the cloud and not on-premises) is completely abstracted away. You don't have to worry about the day-one operations; you only have to worry about getting an application deployed, orchestrated, and working as you and your team see fit.

One important piece of information to think about here is with new technology comes new problems. Kubernetes isn't easy. Just because you don't have to deal with sleeping on a data center floor doesn't mean you won't have an entirely new set of problems to understand and fix. Does Kubernetes make your life easier as an engineer? Yes. Does Kubernetes make your life harder? Yes. Although, the goal is to make your life a little less hard with Kubernetes, please keep in mind that it isn't a magic box that you set and forget.

Cloud-native apps and why they're important

When thinking about creating any type of application, automation code, or piece of software, there always needs to be some sort of standard. The thing is, there are many standards and there isn't a one-size-fits-all solution. Sure, there are (what should be) mandatory standards for writing code such as storing the code in source control and running certain types of tests, but the workflows for each organization will be drastically different.

When it comes to cloud-native applications and applications running on Kubernetes, the thought process of workflows is the same as any other application, but there are true, standard processes that are automatically implemented for you. This includes things such as the following:

- Easy autoscaling
- Self-healing
- Networking out of the box
- And a lot more

In the upcoming section, we'll build on what you learned previously and dive into what cloud-native apps do for organizations.

What cloud-native apps do for organizations

By definition, a cloud-native application gives you the ability to do the following:

- Easily scale
- Make highly available almost out of the box
- Deploy more efficiently
- Continuously make changes in a much easier fashion versus outside of Kubernetes in a bare-metal/data center environment

When thinking about cloud-native applications and the preceding list, microservices typically come to mind. The idea behind microservices, which is a big piece of the idea behind cloud-native, is the ability to make changes faster and more efficiently. When you're dealing with a monolithic application, the application has many dependencies and is essentially tied together. You can't update one piece of the application without bringing down the rest of the application. Blue/green and canary deployments are far more complicated because of the tightly coupled monolithic application. Self-healing and scalability mean scaling the entire application, not just the pieces that need to be scaled, which means more resources (RAM, CPU, and so on) are typically consumed than what's needed.

Cloud-native and the microservice mindset aim to fix this problem. With microservices running inside Kubernetes, there are some extreme benefits. You can manage how many replicas (copies) of the application are running. That way, you can scale them out or scale them back when needed. Self-healing of Pods is far more efficient since if a piece of the application that's running inside of a Pod goes down, it's not a huge deal because it'll come right back up automatically. The applications running inside of Pods, which have one or more containers running inside of the Pods, are loosely coupled, so updating/upgrading versions of the application in a blue/green or canary scenario utilizing a rolling update is far less likely to fail.

When it comes to teams, as in, individual engineers, microservices help a ton. With a monolithic application, there is a fair amount of coordination that has to happen between the team when changing anything in the code base. Although teamwork and communication are crucial, there shouldn't be a reason to let everyone know about a code change in the development environment that you're making to test a piece of functionality without breaking everyone else's code. With how fast organizations want to move in today's world, this process slows engineering teams down to a grinding halt. Not to mention, if an engineer wants to test how the functionality will work with the rest of the application, they shouldn't have to worry about every piece of the application breaking. That's really where microservices shine.

When the Kubernetes architecture was built, it was thought about in the same way as cloud-native applications – a loosely coupled architecture that is easily scalable and doesn't have a ton of dependencies (hence, the microservice movement). Can you run monolithic applications on Kubernetes? Absolutely. Will they still self-heal and autoscale? Absolutely. The idea behind a cloud-native application environment and cloud-native Kubernetes is to use a microservice-style architecture, but you shouldn't let that stop you from jumping into Kubernetes. The primary goal is to have independent services that can be accessed via an **Application Programming Interface (API)**.

The final piece of the puzzle is containerized applications. Before even running an application inside Kubernetes, it must be containerized. When the idea of containers was thought about way before Docker was around, the idea was to have the ability to split an entire application into tiny micro-sized pieces. When containers are built, they're built with the same mindset as the following aspects:

- Self-contained execution environments

- Virtualized OSes

- Microservice architecture with the ability to split up pieces of an entire application and consolidate it into a single container for the ability to easily scale, update, and swap out

The world is cloud-based

One of the worst things that an organization can do in today's world, from an engineering perspective, is to get left behind. The last thing an organization wants is to realize 10 years later that the systems and dependencies that they have in place are so old that no organization or software company is even supporting them anymore. The golden rule before 2015/2016 was to ensure that the architecture and the people/engineers running the architecture were up to date every 5 to 10 years. Now, with how fast technology is moving, it's more like every 2 to 5 years.

When looking at organizations such as Microsoft, Google, and AWS, they're releasing huge changes and updates all the time. When attending a conference such as Microsoft Build or the AWS Summit, the keynotes are filled with game-changing technology with tons of new services coming to the cloud platforms all the time. The reality is that if organizations don't want to be left behind, they can't wait more than 5 years to start thinking about the newest technology.

With that being said, many organizations can't simply upgrade systems every 6 months or every year because they're too large and they don't have enough people to make those migrations and updates. However, technology leaders need to start thinking about what this will look like because the future of the company will be on the line. For example, let's look at the change in Windows Server over the past few years. Microsoft used to constantly talk about new Windows Server versions and features at every conference. Now, it's all about Azure. The technology world is changing drastically.

Where Kubernetes fits in here is that it helps you make cloud-native and fast-moving decisions almost automatically. For example, let's say (in a crazy world) Kubernetes goes away in 3 years. You still have your containerized applications and your code base that's in source control and loosely coupled, which means you can run it anywhere else, such as in a serverless service or even a virtual machine if it comes down to it. With the way that the world is going, it's not necessarily about always using Kubernetes to prevent an organization from going down. It's about what Kubernetes does for engineers, which is that it allows you to manage infrastructure and applications at the API level.

Engineering talent is toward the cloud

One last small piece we will talk about is the future of engineers themselves. New technology professionals are all about learning the latest and greatest. Why? Because they want the ability to stay competitive and get jobs. They want to stay up to date so they can have a long and healthy career. What this means is that they aren't interested in learning about how to run a data center, because the tech world is telling everyone to learn about the cloud.

As time goes on, it's going to become increasingly difficult for organizations to find individuals that can manage and maintain legacy systems. With that being said, there's no end in sight for legacy systems going away. That's why organizations such as banks are still looking for COBOL developers. The thing is, no engineer wants to bet their career in their 20s on learning legacy pieces.

Abstraction is easier, but with a twist

One of the biggest buzzwords in the technology space today is *abstraction*. Abstraction, at its highest level, involves removing certain pieces of work from you that you specifically need to do to get the job done. For example, if a developer needs to run code, they need to run code. They don't need to build virtual machines or deploy networks. They simply need to run an application. Removing the need for a virtual machine or a network is abstracting away what the developer doesn't need to spend time and focus on.

What abstraction does

Let's take a look at what abstraction does from two sides – Dev and Ops.

From a Dev perspective, the goal of a developer is to plan out pieces of an application, write the code to make those pieces work, and deploy them to see how the pieces work together. However, to deploy the code, you used to need a server, an OS, and other components. With platforms such as Kubernetes, developers don't need that anymore. Instead of having to worry about deploying virtual machines, developers simply have to write a Kubernetes manifest that contains a container image with the application inside of it. No more having to worry about day-one operations.

From an Ops perspective, infrastructure engineers or cloud engineers no longer have to worry about having to stop what they're doing to order servers, deploy virtual machines, and fight to make an OS work as expected. Instead, they can write out a Kubernetes manifest and other API-driven techniques (such as IaC) to ensure that a Kubernetes cluster is up and running, operational, and ready to host developer code/container images.

What abstraction doesn't do

One primary thing that abstraction doesn't do is remove the need to think logically and from an architectural perspective for engineering-related work. Abstraction removes what's now considered the *low-hanging fruit* of an environment. For example, a virtual machine with the need to deploy an OS and manage all the components can now be considered low-hanging fruit when the other option is to deploy a Kubernetes cluster and manage the infrastructure at the API level.

The important piece to remember is that engineers and developers still need to think. Abstraction isn't about having a solution where you press a button or two and poof, your application is up and running with scaling and plenty of high availability. Abstraction at this level still requires solid architecture, planning, and repeatable processes.

Start slow and go fast

The final part of this chapter will involve mapping out how you can start slow but, at the same time, go fast when implementing Kubernetes. The idea is that you want to understand what's happening inside of your organization so that you truly know the need for Kubernetes. Once you know that, you can start implementing it as fast as possible without taking on technical debt and management worries. When thinking about how to start slow and go fast, the premise is to understand the *why* behind the conversation around Kubernetes and then once you know that, start iterating.

Understanding the engineering need for Kubernetes

Every good engineer has a lot of goals, but a few of the primary ones are as follows:

- Make my life easier
- Remove the work that isn't important
- Conduct value-driven work for an organization

When it comes to putting out fires, waking up at 2:00 A.M. and rushing around to try to get a server up and running for a developer isn't the most fun part of an engineer's day. Instead, they want to focus on providing value to an organization. Abstraction helps a ton with removing what isn't needed, as does removing toil.

The same goes for developers. They don't want to worry about waiting days or weeks (or longer) to get a server up and running to host an application. They want a quick, efficient, and scalable way to host applications without having to sit around and wait.

The goal is for an engineer to understand the need for Kubernetes. It's easy to look at the latest and greatest technology so that it can be implemented. That's typically the fun part for many engineers, both on the Ops and Dev sides. However, the most important piece is understanding that Kubernetes removes the low-hanging fruit for setting up environments and instead allows you to focus on value-driven work.

Understanding the business need for Kubernetes

There are always two sides to a tech plan in any organization – the technical/engineering side and the business side. On the business side, the primary important pieces are as follows:

- Will Kubernetes help us go faster?
- Will Kubernetes make us more efficient?
- Will Kubernetes help us get to market faster?
- Will Kubernetes help us reduce downtime and engineering overhead?

The answers to those questions are yes and no, and as an engineer, you have to be prepared to answer them. The golden rule is that Kubernetes removes the incredible complexity of racking and stacking a data center, much like the cloud. When talking about Kubernetes to the business, it isn't a conversation around *implementing this Kubernetes thing and all our problems go away*. The conversation is more around *this Kubernetes thing will make our lives easier*.

Planning is the hard part

As engineers, both on the Dev and Ops sides, playing with new technology is fun. Learning new tricks, new platforms, and beefing up your resume to stay competitive in the market is what a lot of individuals think about. Although that's great, you also must think about the *why* behind implementing Kubernetes.

Before moving on to the next chapter, think about these three things:

- Why do I feel like Kubernetes is important?
- How can Kubernetes help my environment progress?
- How can Kubernetes make deploying software easier?

Now, let's summarize what we've learned in this chapter.

Summary

Before you can even think about implementing Kubernetes, you need to learn about what the cloud is doing for engineers, what cloud-native applications are doing for engineers, and why organizations need to start thinking about Kubernetes. This is always the first step in any engineering-related decision since it impacts not only you but the organization as a whole. Because of the way that the tech world is changing, understanding the need for implementing cloud-based solutions and how to move fast but start slow is how organizations have successful Kubernetes deployments and a smooth on-ramp from traditional monolithic applications to implementing microservices.

Now that you know the *why* behind implementing cloud-native technologies such as Kubernetes and what cloud-native applications do for organizations, it's time to start learning about how to get started with Kubernetes. We will start the next chapter by understanding how to implement a Kubernetes service in the top three clouds.

Further reading

To learn more about the topics that were covered in this chapter, take a look at the following resources:

- *Architecting Cloud Computing Solutions*, by Kevin L. Jackson and Scott Goessling: `https://www.packtpub.com/cloud-networking/cloud-computing`

- *vSphere Virtual Machine Management*, by Rebecca Fitzhugh: `https://www.packtpub.com/product/vsphere-virtual-machine-management/9781782172185`

- *Cloud Native Architectures*, by Tom Laszewski, Kamal Arora, Erik Farr, and Piyum Zonooz: `https://www.packtpub.com/product/cloud-native-architectures/9781787280540`

2

Getting the Ball Rolling with Kubernetes and the Top Three Cloud Platforms

When starting your Kubernetes journey, the typical first step is to create a Kubernetes cluster to work with. The reason why is that if you, for example, start by creating a Kubernetes Manifest (more on this in later chapters), you'll have nowhere to deploy the Manifest to because you don't have a Kubernetes cluster. The other reality when it comes to Kubernetes is there's a ton of cloud-native operations management – things such as monitoring a cluster, automating the deployment of a cluster, and scaling a cluster. Because of that, understanding cluster creation is a crucial step in your Kubernetes journey.

In the previous chapter, you learned not only about why Kubernetes is important but also the backstory of why engineers want to use orchestration in today's world. In this chapter, you're going to hit the ground running by creating and managing your very own Kubernetes clusters in the three major clouds – Azure, **Amazon Web Services (AWS)**, and **Google Cloud Platform (GCP)**.

By the end of this chapter, you'll be able to fully create, deploy, manage, and automate Kubernetes clusters running in the three major clouds. The skills that you will pick up from this chapter will also translate across other Kubernetes cluster deployments. For example, you'll be using Terraform to automate the Kubernetes cluster creation, and you can use Terraform to deploy Kubernetes clusters in other clouds and on-premises environments.

In this chapter, we're going to cover the following topics:

- Azure Kubernetes Service
- AWS EKS
- GKE

With each of the topics, you'll learn how to properly run them in a production-level environment. Throughout the rest of the chapter, you'll be working in depth with various amounts of hands-on-driven labs, creating resources automatically and manually.

Technical requirements

For the purpose of this chapter, you should already know how to navigate through each cloud portal and have a general understanding of how you can automate the creation of cloud infrastructure. Although it would be great to dive into those topics in this book, these topics are huge and there are whole books out there dedicated just to them. Because of that, you should know about automated workflows, Terraform, and the cloud prior to getting started.

To work inside the cloud, you will need the following, all of which you can sign up for and get free credit:

- An Azure account
- An AWS account
- A GCP account
- An infrastructure automation tool such as Terraform

The code for this chapter is in the GitHub repository or directory found here: `https://github.com/PacktPublishing/50-Kubernetes-Concepts-Every-DevOps-Engineer-Should-Know/tree/main/Ch2`.

Azure Kubernetes Service

When you're using Microsoft Azure, you have a few options to choose from when using containers and Kubernetes:

- **Azure Kubernetes Service (AKS)**
- Azure Container Instances
- **Azure Container Apps (ACA)**

AKS is the primary way to run Kubernetes workloads inside Azure. You do not have to worry about managing the Control Plane or API Server and instead, simply handle deploying your apps, scaling, and managing or maintaining the cloud infrastructure. However, there is still maintenance and management that you need to do for worker nodes – for example, if you want to scale Kubernetes clusters, utilize a multi-cloud model, or implement some sort of hybrid-cloud model, you would be solely responsible for implementing that setup. AKS abstracts the need to worry about managing and scaling the Control Plane or API Server, but you're responsible for everything else (scaling workloads, monitoring, and observability).

> **Important note**
>
> There's a new service that recently went **Generally Available (GA)** at Microsoft Build 2022 called ACA. Although we won't be going into detail about ACA in this book, you should know that it's essentially *serverless Kubernetes*. It's drastically different in comparison to AKS, so if you're planning on using ACA, ensure that you learn about those tech spaces prior.

In the following section, you're going to learn how to create an AKS cluster manually first. After that, you'll take what you learned from a manual perspective and learn how to automate it with Terraform. Then, you'll learn about scaling AKS clusters from a vertical-autoscaling perspective. Finally, you'll wrap up with serverless Kubernetes. Let's dive right in!

Creating an AKS cluster manually

Before managing an AKS cluster, you have to learn how to create one. In today's world, you'll most likely never do this process manually because of the need for every organization to strive toward an automated and repeatable mindset. However, because you cannot automate something without doing it manually first, you'll learn how to do that in this section:

1. Log in to the Azure portal.

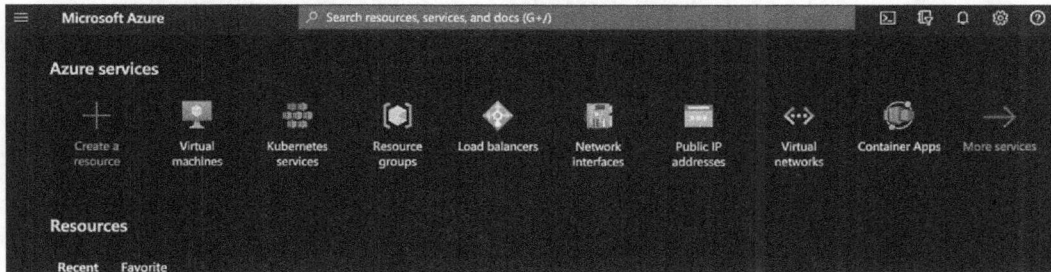

Figure 2.1 – The Azure portal

2. Search for `azure kubernetes services`:

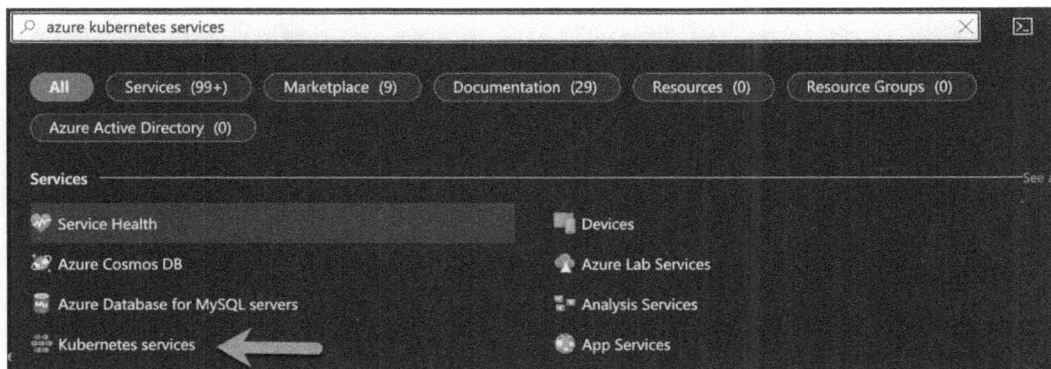

Figure 2.2 – Searching for AKS

3. Click on the **Create** dropdown and choose the **Create a Kubernetes cluster** option:

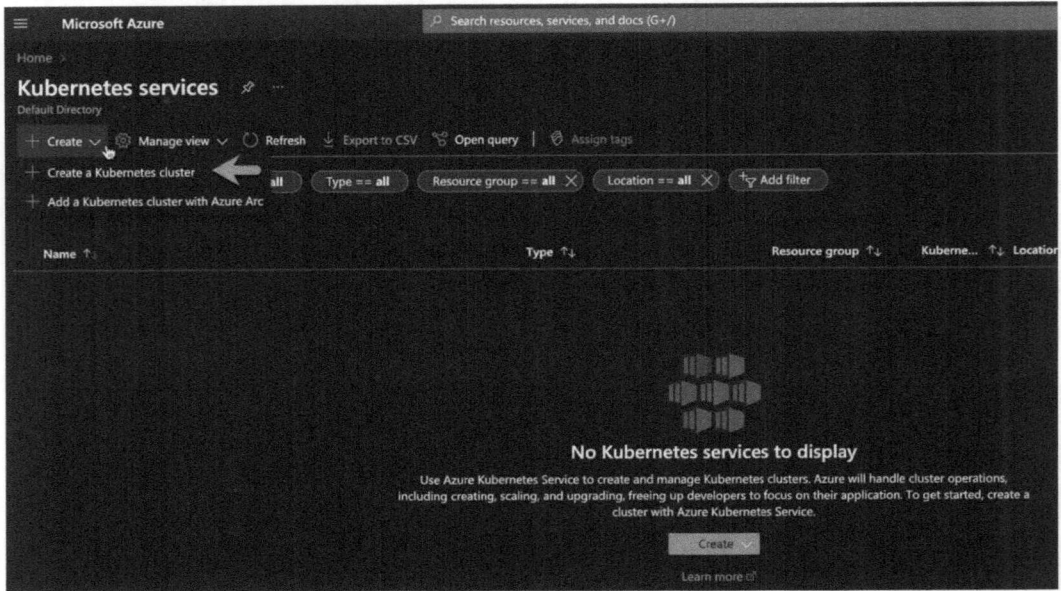

Figure 2.3 – Creating an AKS cluster

4. Choose the options for your Kubernetes cluster, including the name of the cluster and your Azure resource group:

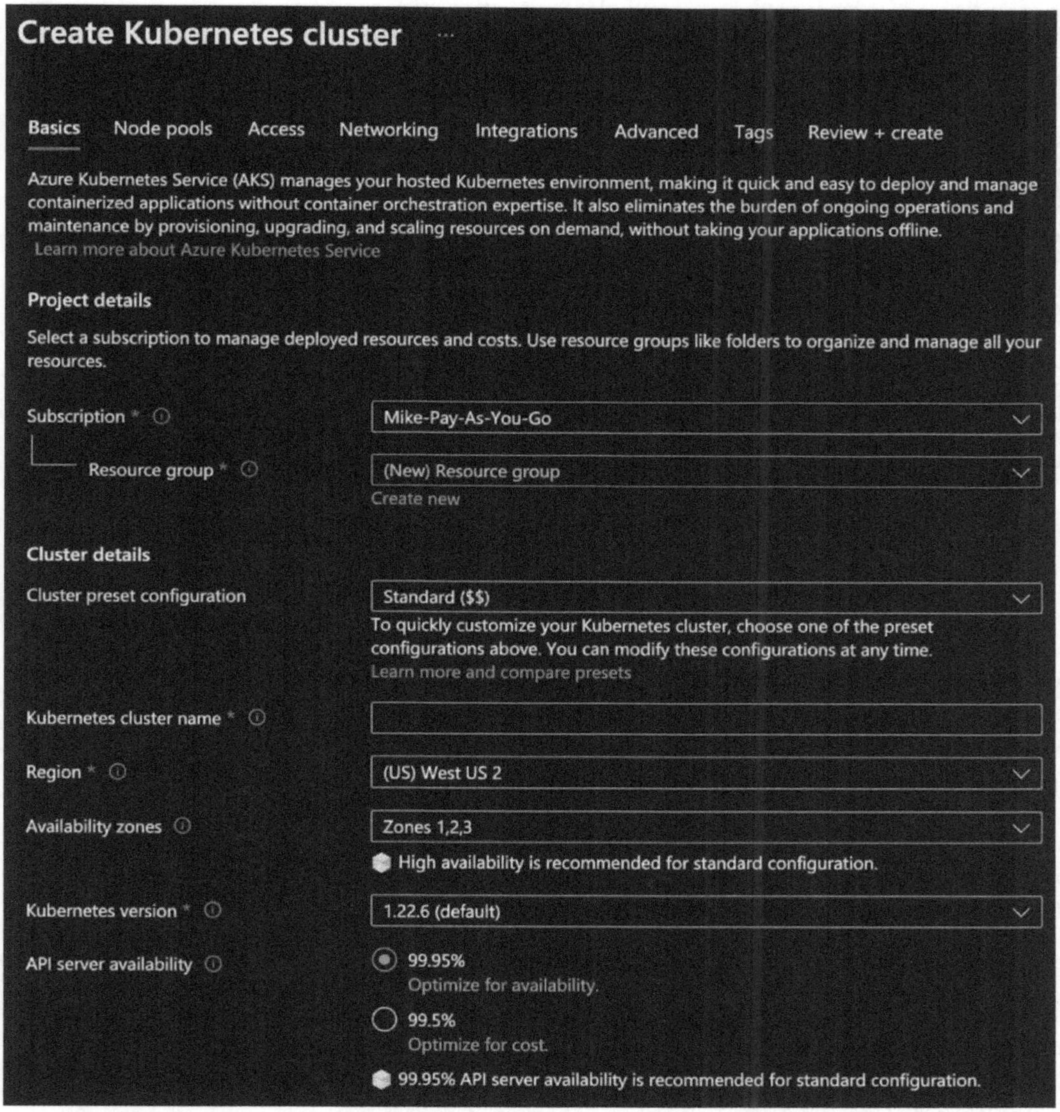

Create Kubernetes cluster ...

Basics   Node pools   Access   Networking   Integrations   Advanced   Tags   Review + create

Azure Kubernetes Service (AKS) manages your hosted Kubernetes environment, making it quick and easy to deploy and manage containerized applications without container orchestration expertise. It also eliminates the burden of ongoing operations and maintenance by provisioning, upgrading, and scaling resources on demand, without taking your applications offline.
Learn more about Azure Kubernetes Service

Project details

Select a subscription to manage deployed resources and costs. Use resource groups like folders to organize and manage all your resources.

Subscription * ⓘ

| Mike-Pay-As-You-Go | ⌄ |

└─── Resource group * ⓘ

| (New) Resource group | ⌄ |
Create new

Cluster details

Cluster preset configuration

| Standard ($$) | ⌄ |

To quickly customize your Kubernetes cluster, choose one of the preset configurations above. You can modify these configurations at any time.
Learn more and compare presets

Kubernetes cluster name * ⓘ

| |

Region * ⓘ

| (US) West US 2 | ⌄ |

Availability zones ⓘ

| Zones 1,2,3 | ⌄ |

🔵 High availability is recommended for standard configuration.

Kubernetes version * ⓘ

| 1.22.6 (default) | ⌄ |

API server availability ⓘ

◉ 99.95%
Optimize for availability.

○ 99.5%
Optimize for cost.

⬡ 99.95% API server availability is recommended for standard configuration.

Figure 2.4 – Adding cluster details before its creation

5.   Under the **Primary node pool** section, you can choose what **Virtual Machine** (**VM**) size you want for your Kubernetes worker nodes, how many should be available, and whether or not you want to autoscale. One of the biggest powers behind cloud Kubernetes services such as AKS is autoscaling. In production, it's recommended to autoscale when needed. However, you also have to understand that it comes with a cost, as extra VMs will be provisioned. Leave everything as the default for now and scroll down to the **Primary node pool** section:

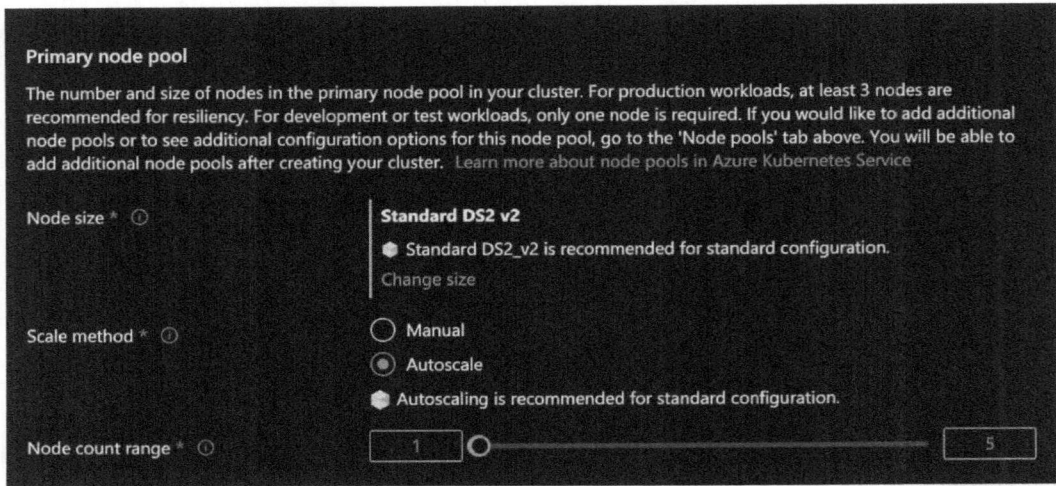

Figure 2.5 – Specifying the worker node size, node count, and scale method

6. Once you have chosen your options, which in a dev environment could be one node, but will vary in production, click the blue **Review + create** button. Your AKS cluster is now created.

Now that you know how to create an AKS cluster manually, it's time to learn how to create it with Terraform so you can ensure repeatable processes throughout your environment.

Creating an AKS cluster with automation

In many production-level cases, you'll run the following Terraform code within a CI/CD pipeline to ensure repeatability. For the purpose of this section, you can run it locally. You'll first see the main. tf configuration and then you'll take a look at variables.tf.

Let's break down the code.

First, there's the Terraform provider itself. The azurerm Terraform provider is used to make API calls to Azure programmatically:

```
terraform {
  required_providers {
    azurerm = {
      source = "hashicorp/azurerm"
    }
  }
}
```

```
provider "azurerm" {
  features {}
}
```

Next, there's the `azurerm_kubernetes_cluster` Terraform resource block, which is used to create the AKS cluster. There are a few key parameters, including the name and **Domain Name System (DNS) Fully Qualified Domain Name (FQDN)**. The Kubernetes worker nodes are created via the `default_node_pool` parameter block. You can specify the VM size, node count, and name of the node pool:

```
resource "azurerm_kubernetes_cluster" "k8squickstart" {
  name                = var.name
  location            = var.location
  resource_group_name = var.resource_group_name
  dns_prefix          = "${var.name}-dns01"

  default_node_pool {
    name       = "default"
    node_count = var.node_count
    vm_size    = "Standard_A2_v2"
  }

  identity {
    type = "SystemAssigned"
  }

  tags = {
    Environment = "Production"
  }
}
```

Putting it all together, you'll have a Terraform configuration that creates an AKS cluster in Azure.

Now that you have the Terraform configuration, you'll need variables to pass in. The variables allow your code to stay repeatable – in accordance with the **Don't Repeat Yourself (DRY)** principle – so that you don't have to continuously change hardcoded values or create new configurations for each environment.

There are four variables:

- `name`: Name of the AKS cluster
- `resource_group_name`: The resource group that AKS will reside in
- `location`: The region that the AKS cluster will reside in
- `node_count`: How many Kubernetes worker nodes will be in the AKS cluster

These variables can be seen in the following code block:

```
variable "name" {
  type = string
  default = "aksenvironment01"
}

variable "resource_group_name" {
  type = string
  default = "devrelasaservice"
}

variable "location" {
  type = string
  default = "eastus"
}

variable "node_count" {
  type = string
  default = 3
}
```

Putting both the `main.tf` and `variables.tf` configuration files in the same directory will create a Terraform module for creating an AKS cluster. You can use this for almost any environment, change configurations (such as the node count) depending on your needs, and make your process repeatable.

Scaling an AKS cluster

Scaling an AKS cluster is made possible by implementing the Kubernetes Cluster Autoscaler. Much like autoscaling groups for Azure VMs, AKS decides on how and why to scale the cluster based on the worker node load, which is the Azure VMs in the background. The Cluster Autoscaler is typically deployed to the Kubernetes cluster with the `cluster-autoscaler` container image.

Log in to the Azure portal and go to the AKS service. Once there, go to **Settings | Node pools**:

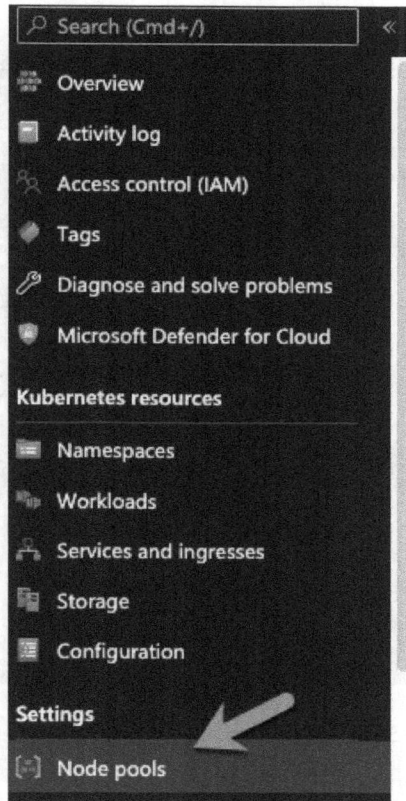

Figure 2.6 – Node pools settings

Click on the three dots as per the following screenshot and choose the **Scale node pool** option:

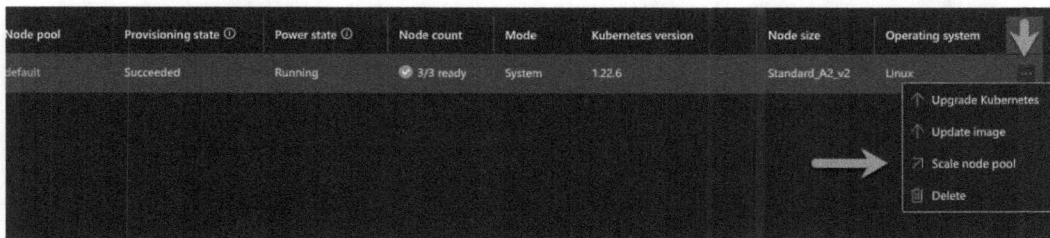

Figure 2.7 – Scaling node pools

The **Scale node pool** pane will come up and you'll see the option to automatically scale the node pool or manually scale it and choose how many nodes you want to make available:

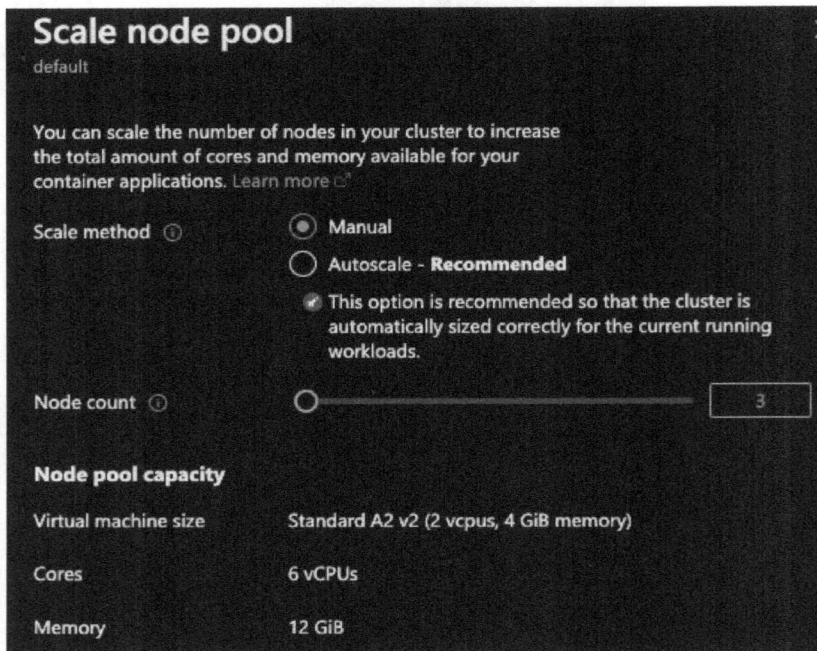

Figure 2.8 – Specifying the node count and scale method

From an automation and repeatability standpoint, you can do the same thing. The following is an example of creating the `azurerm_kubernetes_cluster_node_pool` Terraform resource with the `enable_auto_scaling` parameter set to `true`:

```
resource "azurerm_kubernetes_cluster_node_pool" "example" {
  name                  = "internal"
  kubernetes_cluster_id = azurerm_kubernetes_cluster.example.id
  vm_size               = "Standard_DS2_v2"
  node_count            = 1
  enable_auto_scaling   = true

  tags = {
    Environment = "Production"
  }
}
```

Node pools are simply Azure VMs that run as Kubernetes worker nodes. When thinking about autoscaling, remember that horizontal autoscaling comes at a cost. Although it's very much needed, you should limit the amount of Kubernetes worker nodes that are available. That way, you can keep track of costs and how many resources your containerized apps need.

AKS and Virtual Kubelet

To wrap things up with AKS, there is Virtual Kubelet. Virtual Kubelet isn't AKS-specific. Virtual Kubelet allows you to take Kubernetes and connect it to other APIs. A kubelet is the node agent that runs on each node. It's responsible for registering the node with Kubernetes. AKS Virtual Kubelet registers serverless container platforms.

In Azure, it's **Azure Container Instances** (**ACI**). ACI is a way to run containers without using Kubernetes. If someone using Kubernetes doesn't want to scale out worker nodes due to cost or management, they can use ACI bursting, which uses Virtual Kubelet. It essentially tells Kubernetes to send Deployments, Pods, and other workloads to ACI instead of keeping them on the local Kubernetes cluster.

Now that ACA is GA, chances are you'll see less of this type of implementation. However, it's still a great use case for teams that want to scale, but don't want the overhead of managing large AKS clusters.

Managing and maintaining AKS clusters

Once a Kubernetes cluster is created and running, the mindset shift moves from day-one Ops to day-two Ops. Day-two Ops will be focused on the following:

- Managing the cluster
- Monitoring and maintaining the cluster
- Deploying applications and getting services running

When managing an AKS cluster, the biggest thing to think about is where the configuration exists and what tools you're using to manage it. For example, the Terraform configuration could be in GitHub and you could be managing the cluster via Azure Monitor and the rest of the Azure configurations that are available in the AKS cluster. Day-two Ops is about making sure the cluster and your configurations are running as you expect. The focus is really on the question "*is my environment working and performing as intended?*"

When it comes to monitoring, alerting, and overall observability, there are several options. Azure Monitor and Azure Insights are built into Azure, but if you have a multi-cloud or a hybrid-cloud environment, you may want to look at other options. That's where a combination of Prometheus and Grafana can come into play. Whichever tool you choose (because there are several) isn't important. What's important is what you monitor. You'll need a combination of monitoring the cluster itself and the Kubernetes resources (for example, Pods, Services, or Ingresses) inside of the cluster.

Because managing Kubernetes clusters doesn't differ all that much (other than the native cloud tools), it's safe to assume that whether you're using AKS, EKS, or GKE, the path forward will be the same.

AWS EKS

When you're using AWS, you have a few options to choose from when using containers and Kubernetes:

- EKS
- EKS with Fargate profiles
- **Elastic Container Service (ECS)**

EKS is the primary way to run Kubernetes workloads inside AWS. If you don't want to go the Kubernetes route but still want scalability, you can use ECS, which gives you the ability to scale and create reliable microservices but without Kubernetes.

As with AKS, you don't have to worry about managing the Control Plane or API Server when it comes to EKS. You only have to worry about managing and scaling worker nodes. If you want to, you can even take it a step further and implement EKS with Fargate profiles, which abstracts the Control Plane or API Server and the worker nodes to ensure a fully *serverless Kubernetes* experience.

As with AKS, in the following few sections, you're going to learn how to create an EKS cluster manually first. After that, you'll take what you learned from a manual perspective and learn how to automate it with Terraform. Then, you'll learn about scaling EKS clusters from a vertical-autoscaling perspective. Finally, you'll wrap up with serverless Kubernetes.

Creating an EKS cluster manually

Much like AKS clusters, before creating EKS clusters from an automated perspective, you must learn how to manually deploy them. In this section, you'll learn how to deploy an EKS cluster with a node group in the AWS Console:

1. Log in to the AWS portal and search for the EKS service:

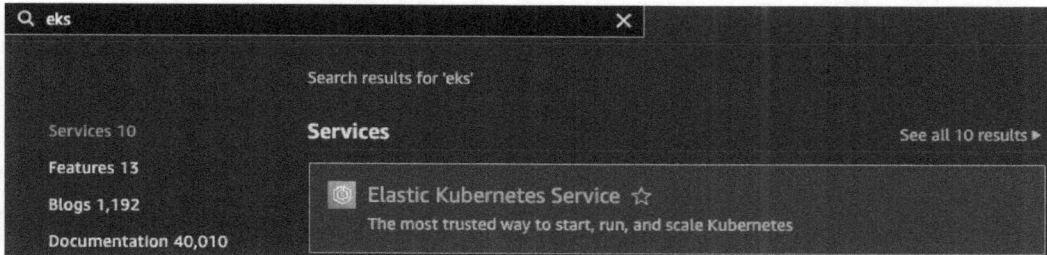

Figure 2.9 – The AWS portal

2. Click the orange **Add cluster** button and choose the **Create** option:

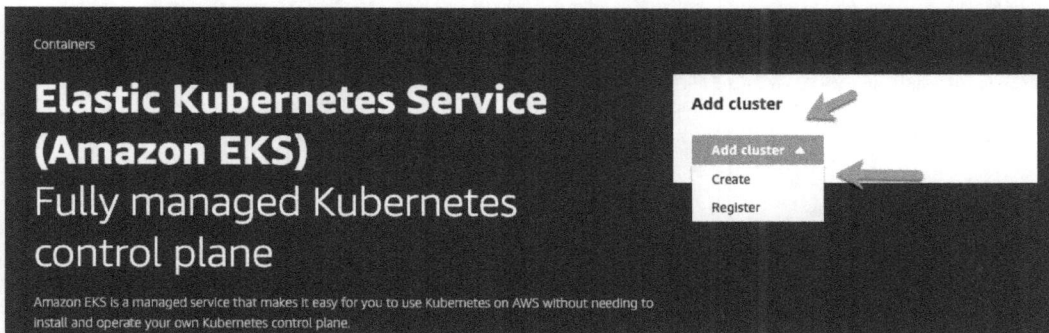

Figure 2.10 – Adding a cluster

3. When configuring an EKS cluster, you'll have to provide a few options to uniquely identify it, which include the following:

    -   The EKS cluster name

    -   The Kubernetes API version

    -   The IAM role

    The IAM role is very important because there are specific policies that must be attached to the role that you're assigning to the EKS cluster. Those policies include the following:

    -   `AmazonEC2ContainerRegistryReadOnly`

    -   `AmazonEKSClusterPolicy`

Without the proceeding policies, the EKS cluster will not work as expected:

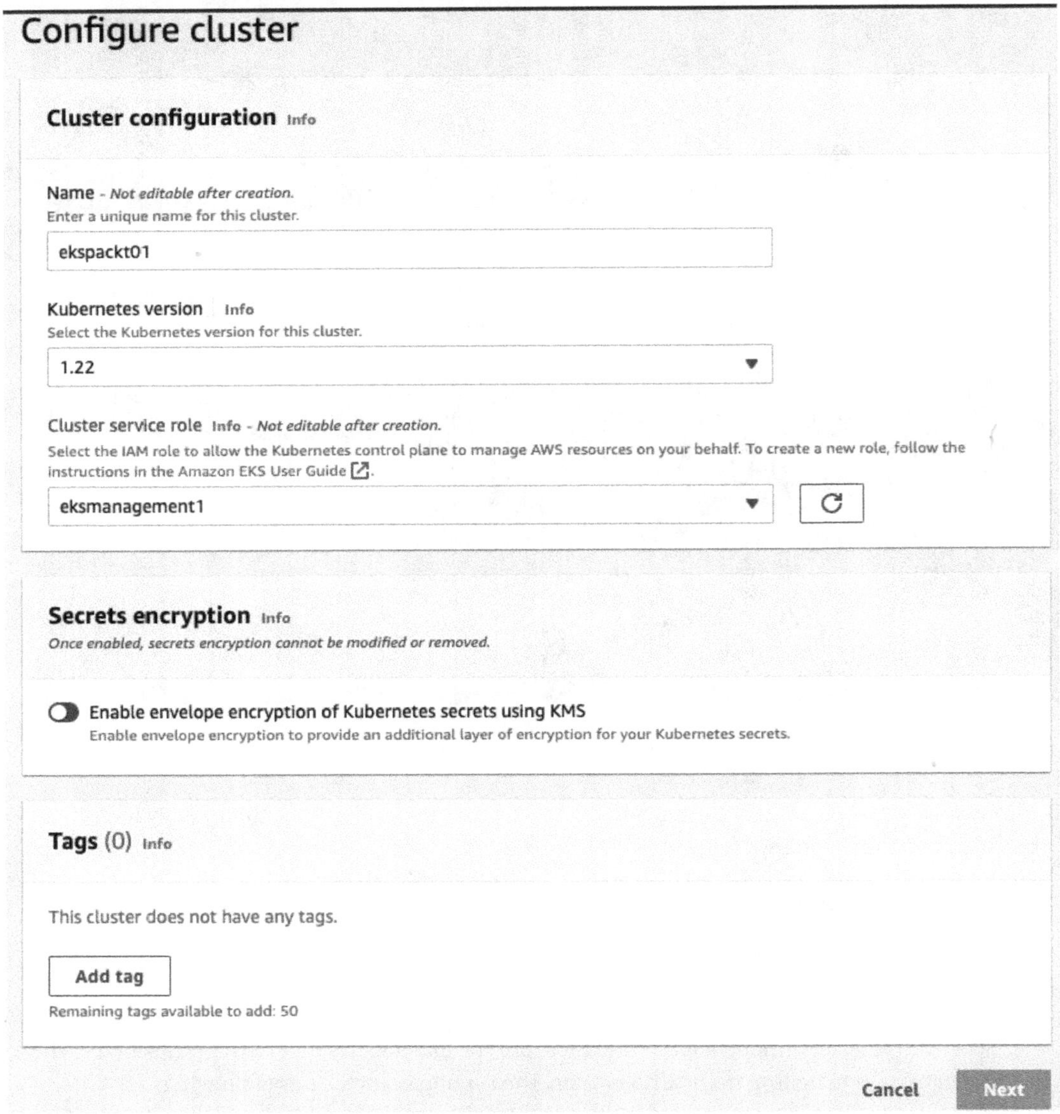

Configure cluster

Cluster configuration Info

Name - *Not editable after creation.*
Enter a unique name for this cluster.

> ekspackt01

Kubernetes version Info
Select the Kubernetes version for this cluster.

> 1.22   ▼

Cluster service role Info - *Not editable after creation.*
Select the IAM role to allow the Kubernetes control plane to manage AWS resources on your behalf. To create a new role, follow the instructions in the Amazon EKS User Guide [↗].

> eksmanagement1   ▼ | C |

Secrets encryption Info
Once enabled, secrets encryption cannot be modified or removed.

⬤ **Enable envelope encryption of Kubernetes secrets using KMS**
    Enable envelope encryption to provide an additional layer of encryption for your Kubernetes secrets.

Tags (0) Info

This cluster does not have any tags.

> **Add tag**

Remaining tags available to add: 50

Cancel   Next

Figure 2.11 – Configuring a cluster

4. Next, you'll need to set up networking. The absolute minimum amount of subnets that you want to use is two public subnets with different CIDRs in different availability zones. For a full list of recommendations, check out the AWS docs (`https://docs.aws.amazon.com/eks/latest/userguide/network_reqs.html`):

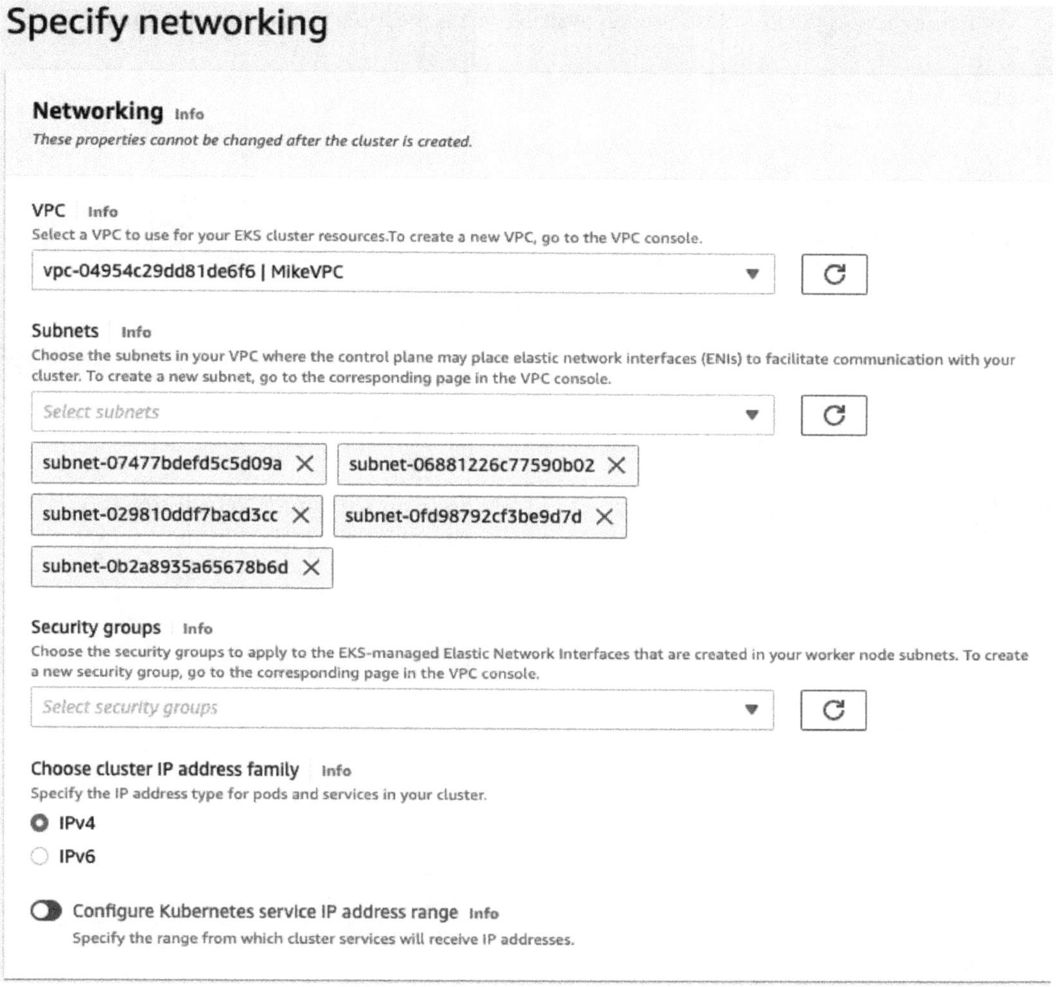

Figure 2.12 – Specifying the network configuration

5. When configuring the cluster endpoint access, you have three options:

    • **Public** means the EKS cluster is essentially open to the world

    • **Public and private** means the API Server or Control Plane is open to the outside world, but worker node traffic will remain internal

- **Private** means the EKS cluster is only available inside the AWS **Virtual Private Cloud** (**VPC**):

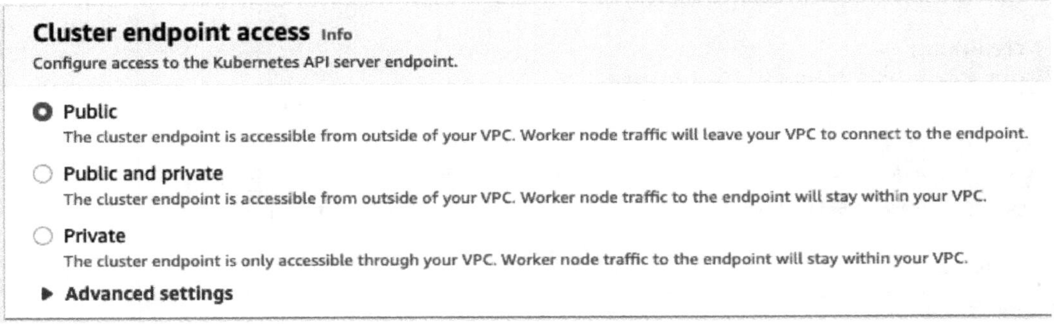

Cluster endpoint access Info
Configure access to the Kubernetes API server endpoint.

○ **Public**
The cluster endpoint is accessible from outside of your VPC. Worker node traffic will leave your VPC to connect to the endpoint.

○ **Public and private**
The cluster endpoint is accessible from outside of your VPC. Worker node traffic to the endpoint will stay within your VPC.

○ **Private**
The cluster endpoint is only accessible through your VPC. Worker node traffic to the endpoint will stay within your VPC.

▶ **Advanced settings**

Figure 2.13 – Configuring cluster API access

6. The last piece from a networking perspective is choosing the **Container Networking Interface** (**CNI**) and the version of CoreDNS. Choosing the latest typically makes the most sense:

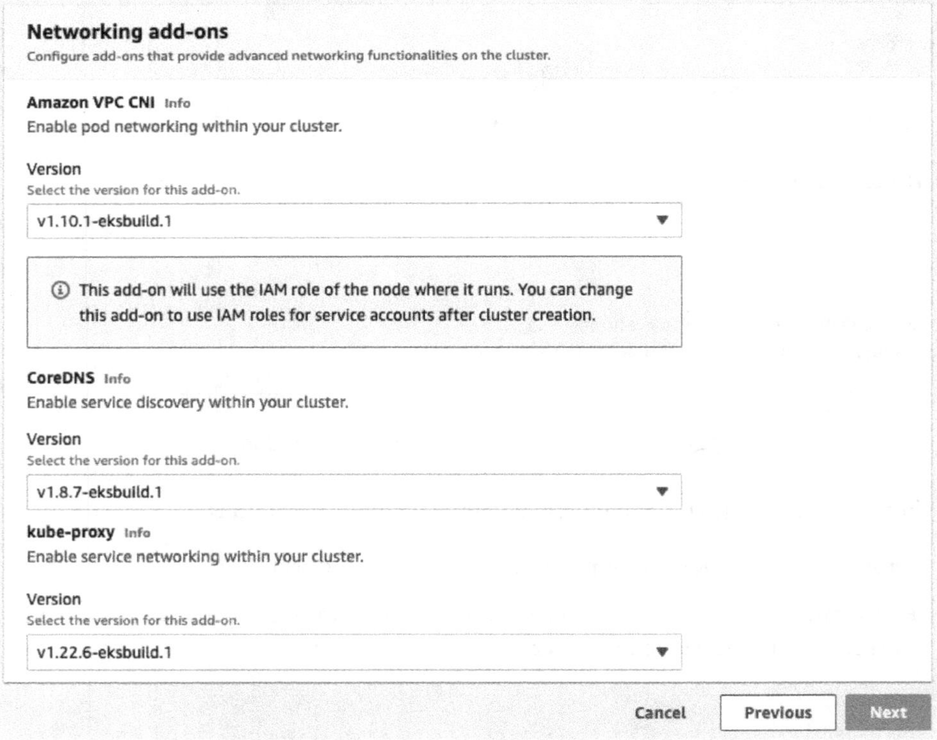

Networking add-ons
Configure add-ons that provide advanced networking functionalities on the cluster.

Amazon VPC CNI Info
Enable pod networking within your cluster.

Version
Select the version for this add-on.

| v1.10.1-eksbuild.1 | ▼ |

ⓘ This add-on will use the IAM role of the node where it runs. You can change this add-on to use IAM roles for service accounts after cluster creation.

CoreDNS Info
Enable service discovery within your cluster.

Version
Select the version for this add-on.

| v1.8.7-eksbuild.1 | ▼ |

kube-proxy Info
Enable service networking within your cluster.

Version
Select the version for this add-on.

| v1.22.6-eksbuild.1 | ▼ |

Cancel   Previous   Next

Figure 2.14 – Network add-ons

7. Click the orange **Next** button.

8. The final piece when creating the EKS cluster is the API logging. Regardless of where you plan to keep logs, traces, and metrics from an observability perspective, you must turn this option *on* if you want your cluster to record any type of logs:

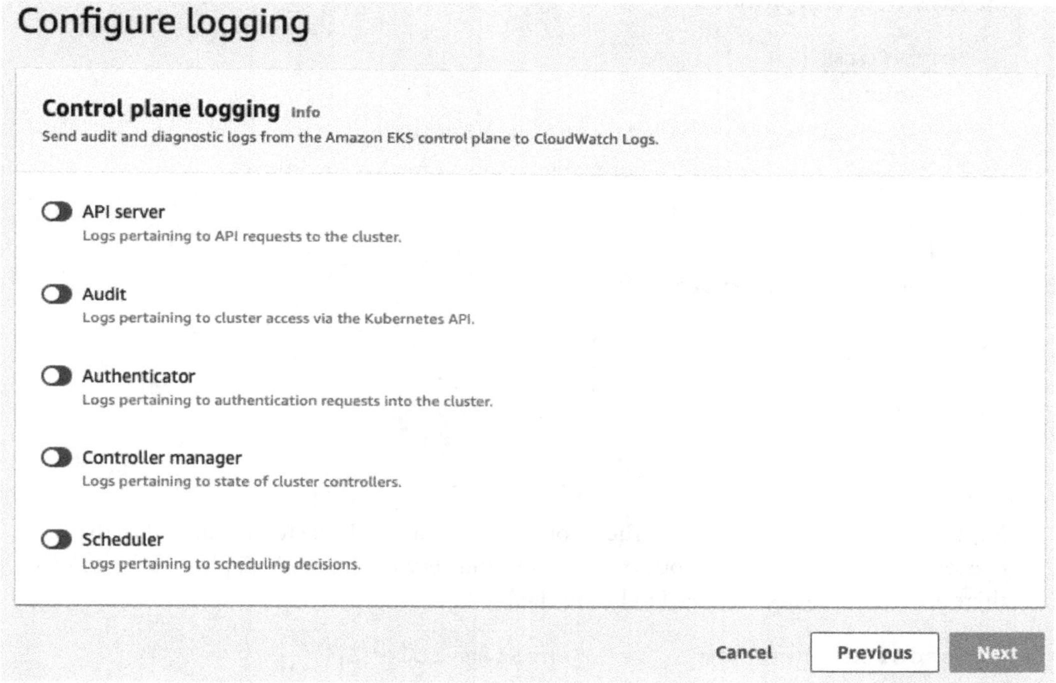

Configure logging

Control plane logging Info
Send audit and diagnostic logs from the Amazon EKS control plane to CloudWatch Logs.

API server
Logs pertaining to API requests to the cluster.

Audit
Logs pertaining to cluster access via the Kubernetes API.

Authenticator
Logs pertaining to authentication requests into the cluster.

Controller manager
Logs pertaining to state of cluster controllers.

Scheduler
Logs pertaining to scheduling decisions.

Cancel   Previous   Next

Figure 2.15 – Configuring observability

9. After you choose your logging options, click the orange **Next** button and you'll be at the last page to review and create your EKS cluster.

Now that you know how to create an EKS cluster manually, it's time to learn how to create it with Terraform so you can ensure repeatable processes throughout your environment.

Creating an EKS cluster with Terraform

In many production-level cases, you'll run the following Terraform code within a CI/CD pipeline to ensure repeatability. For the purposes of this section, you can run it locally.

First, you'll see the `main.tf` configuration and then you'll take a look at `variables.tf`.

Because the `main.tf` configuration for AWS EKS is much longer than EKS, let's break it down into chunks for an easier explanation:

1. First, there's the Terraform provider block. To ensure repeatability throughout your team, you can use an S3 bucket backend for storing your `TFSTATE`. The Terraform block also includes the AWS Terraform provider:

```
terraform {
  backend "s3" {
    bucket = "terraform-state-k8senv"
    key    = "eks-terraform-workernodes.tfstate"
    region = "us-east-1"
  }
  required_providers {
    aws = {
      source = "hashicorp/aws"
    }
  }
}
```

2. Next, the first resource is created. The resources allow an IAM role to be attached to the EKS cluster. For EKS to access various components and services from AWS, plus worker nodes, there are a few policies that need to be attached:

```
resource "aws_iam_role" "eks-iam-role" {
  name = "k8squickstart-eks-iam-role"

  path = "/"

  assume_role_policy = <<EOF
{
  "Version": "2012-10-17",
  "Statement": [
    {
      "Effect": "Allow",
      "Principal": {
        "Service": "eks.amazonaws.com"
      },
      "Action": "sts:AssumeRole"
```

```
    }
  ]
}
EOF

}
```

3. Following the IAM role are the IAM policies that have to be attached to the role. The two policies that you'll need for a successful EKS deployment are the following:

- `AmazonEKSClusterPolicy`: This provides Kubernetes with the permissions it requires to manage resources on your behalf:

```
resource "aws_iam_role_policy_attachment"
"AmazonEKSClusterPolicy" {
  policy_arn = "arn:aws:iam::aws:policy/
AmazonEKSClusterPolicy"
  role       = aws_iam_role.eks-iam-role.name
}
```

- `AmazonEC2ContainerRegistryReadOnly`: This provides read-only access to Elastic Container Registry if you decide to put your container images there:

```
resource "aws_iam_role_policy_attachment" "AmazonEC2Conta
inerRegistryReadOnly-EKS" {
  policy_arn = "arn:aws:iam::aws:policy/
AmazonEC2ContainerRegistryReadOnly"
  role       = aws_iam_role.eks-iam-role.name
}
```

4. Once the IAM role and policies are defined, it's time to create the EKS cluster itself. The EKS cluster resource will create EKS itself, enable logging, and attach the IAM role that you created earlier:

```
resource "aws_eks_cluster" "k8squickstart-eks" {
  name = "k8squickstart-cluster"
  role_arn = aws_iam_role.eks-iam-role.arn

  enabled_cluster_log_types = ["api", "audit",
"scheduler", "controllerManager"]

  vpc_config {
```

```
        subnet_ids = [var.subnet_id_1, var.subnet_id_2]
    }

    depends_on = [
      aws_iam_role.eks-iam-role,
    ]
}
```

5. The next resource is another IAM role, which is for the worker nodes. When creating an EKS cluster, you'll have multiple resources that are created because you're creating two sets of services:

    -   The EKS cluster itself with all of its permissions and policies that are needed

    -   The Kubernetes worker nodes with all of the permissions and policies needed:

        ```
        resource "aws_iam_role" "workernodes" {
          name = "eks-node-group-example"

          assume_role_policy = jsonencode({
            Statement = [{
              Action = "sts:AssumeRole"
              Effect = "Allow"
              Principal = {
                Service = "ec2.amazonaws.com"
              }
            }]
            Version = "2012-10-17"
          })
        }
        ```

6. Once the IAM role for the worker nodes is created, there are a few policies that you'll need to attach:

    -   `AmazonEKSWorkerNodePolicy`: This provides Kubernetes the permissions it requires to manage resources on your behalf:

        ```
        resource "aws_iam_role_policy_attachment"
        "AmazonEKSWorkerNodePolicy" {
          policy_arn = "arn:aws:iam::aws:policy/
        AmazonEKSWorkerNodePolicy"
          role       = aws_iam_role.workernodes.name
        }
        ```

- `AmazonEKS_CNI_Policy`: This attaches the CNI policy for Kubernetes internal networking (kubeproxy):

```
resource "aws_iam_role_policy_attachment" "AmazonEKS_CNI_
Policy" {
  policy_arn = "arn:aws:iam::aws:policy/AmazonEKS_CNI_
Policy"
  role       = aws_iam_role.workernodes.name
}
```

- `EC2InstanceProfileForImageBuilderECRContainerBuilds`: EC2 Image Builder uses a service-linked role to grant permissions to other AWS services on your behalf:

```
resource "aws_iam_role_policy_attachment"
"EC2InstanceProfileForImageBuilderECRContainerBuilds" {
  policy_arn = "arn:aws:iam::aws:policy/
EC2InstanceProfileForImageBuilderECRContainerBuilds"
  role       = aws_iam_role.workernodes.name
}
```

- `AmazonEC2ContainerRegistryReadOnly`: This provides read-only access to Elastic Container Registry if you decide to put your container images there:

```
resource "aws_iam_role_policy_attachment"
"AmazonEC2ContainerRegistryReadOnly" {
  policy_arn = "arn:aws:iam::aws:policy/
AmazonEC2ContainerRegistryReadOnly"
  role       = aws_iam_role.workernodes.name
}
```

- `CloudWatchAgentServerPolicy`: This allows the worker nodes to run the CloudWatch agent for monitoring, logging, tracing, and metrics:

```
resource "aws_iam_role_policy_attachment"
"CloudWatchAgentServerPolicy-eks" {
  policy_arn = "arn:aws:iam::aws:policy/
CloudWatchAgentServerPolicy"
  role       = aws_iam_role.workernodes.name
}
```

7. The final part once the IAM role and policies have been created is to create the EKS node group resource, which is the Kubernetes worker nodes. You'll define the following:

- The IAM role and subnet IDs:

```
resource "aws_eks_node_group" "worker-node-group" {
  cluster_name      = aws_eks_cluster.k8squickstart-eks.
name
  node_group_name = "k8squickstart-workernodes"
  node_role_arn   = aws_iam_role.workernodes.arn
  subnet_ids      = [var.subnet_id_1, var.subnet_id_2]
  instance_types = ["t3.xlarge"]
```

- The desired scale size for autoscaling:

```
scaling_config {
  desired_size = 3
  max_size     = 4
  min_size     = 2
}
```

- The policies that the resource depends on:

```
depends_on = [
  aws_iam_role_policy_attachment.
AmazonEKSWorkerNodePolicy,
  aws_iam_role_policy_attachment.AmazonEKS_CNI_Policy,

  ]
}
```

8. Now that you have the Terraform configuration, you'll need variables to pass in. The variables allow your code to stay repeatable, so you don't have to continuously change hardcoded values or create new configurations for each environment.

The two variables you'll need are for the subnet IDs in the VPC of your choosing that will work with EKS. You can pass in two public subnet IDs that are in different availability zones:

```
variable "subnet_id_1" {
  type = string
  default = ""
```

```
}

variable "subnet_id_2" {
  type = string
  default = ""
}
```

Putting it all together, you'll have a Terraform configuration that creates an AKS cluster.

Scaling an EKS cluster

Scaling an EKS cluster is made possible by implementing the Kubernetes Cluster Autoscaler. Much like autoscaling EC2 instances, EKS decides on how and why to scale the cluster based on a load perspective. The Cluster Autoscaler is typically deployed to the Kubernetes cluster using the `cluster-autoscaler` container image.

Inside the Kubernetes GitHub repo, under the `cluster-autoscaler` directory, there's a list of cloud providers. One of those cloud providers is AWS. Inside the AWS directory, there's an example Kubernetes Manifest called `cluster-autoscaler-autodiscover.yaml`, which shows that it's using the `cluster-autoscaler` container image. It runs as a Kubernetes Deployment on your cluster and listens for certain resource limits. To autoscale the cluster, you'll need an IAM role with the `AmazonEKSClusterAutoscalerPolicy` policy attached to it.

Now that you know about scaling an EKS cluster and how it's possible with `cluster-autoscaler`, let's talk about serverless Kubernetes with AWS Fargate profiles and how they can help automate day-one Ops.

EKS Fargate profiles

The content around Fargate profiles is pretty similar to AKS Virtual Kubelet and ACI bursting. However, you don't need to deploy Virtual Kubelet manually as you do in AKS. Instead, you can set up Fargate profiles to act as your Kubernetes worker nodes. Virtual Kubelet is still running on Fargate to interact with the EKS API Server or Control Plane, but it's sort of done automatically.

The biggest difference here is that you don't have to manage the worker nodes. Instead, Fargate profiles are like serverless Kubernetes. You deploy the EKS cluster, which is the API Server or Control Plane. Then, you deploy a Fargate profile, which is where your Kubernetes resources (for example, Deployments, Pods, and Services) run. You don't have to worry about cluster management or maintaining EC2 instances that would otherwise be running as your Kubernetes worker nodes.

To add a Fargate profile on your EKS cluster, you go into the **Compute** tab of the EKS cluster and you'll see an option for adding or creating a Fargate profile, as seen in the following screenshot:

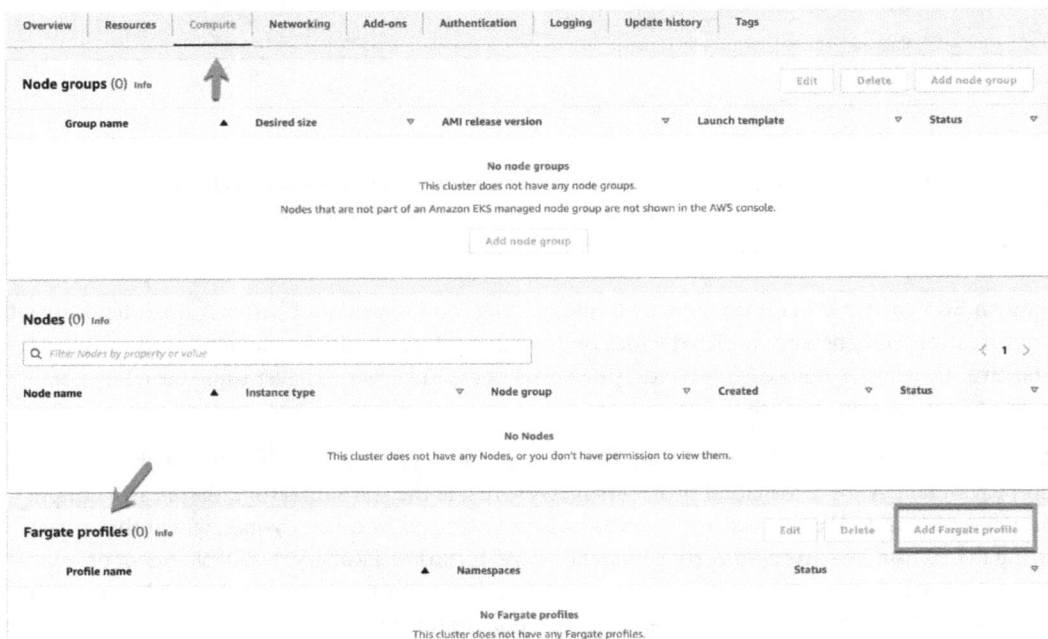

Figure 2.16 – Fargate profiles and compute

Now that you know how to create an EKS cluster manually and automatically, and are also familiar with the day-two Ops considerations with autoscaling and serverless Kubernetes, it's time to learn about the final *big 3* Kubernetes service – GKE.

GKE

When you're using GCP, you have a few options to choose from when using containers and Kubernetes:

- GKE
- GKE Autopilot
- Google Cloud Run

GKE is the primary way to run Kubernetes workloads inside of GCP. If you don't want to go the Kubernetes route but still want scalability, you can use Google Cloud Run. Cloud Run gives you the ability to scale and create reliable microservices, but without Kubernetes. It supports Node.js, Go, Java, Kotlin, Scala, Python, .NET, and Docker.

As with AKS and EKS, you don't have to worry about managing the Control Plane or API Server when it comes to GKE. You only have to worry about managing and scaling worker nodes. If you want to, you can even take it a step further and implement GKE Autopilot, which abstracts both the Control Plane or API Server and the worker nodes to ensure a fully *serverless Kubernetes* experience.

There have been many debates inside of container and DevOps communities around which Kubernetes service in the cloud is the superior choice. Although we're not here to pick sides, a lot of engineers love GKE and believe it's a spectacular way to implement Kubernetes. Since Kubernetes originated at Google, it makes sense that the GKE service would be incredibly reliable with well-thought-out features and implementations.

In the following section, you will learn about creating a GKE cluster automatically using Terraform and how to think about serverless Kubernetes using GKE Autopilot.

> **Important note**
> We're skipping a section on scaling the GKE cluster because it's the same concept as the other clouds. It uses the Kubernetes Autoscaler in the background. All the autoscalers are considered horizontal autoscalers, as they create new worker nodes or VMs to run Kubernetes workloads.

Creating a GKE cluster with Terraform

Throughout this chapter, you've learned several manual ways of creating a Kubernetes cluster in the cloud. Instead of continuing down the manual road, let's jump right into automating the repeatable process of creating a GKE cluster with Terraform.

What you'll find with GKE is that it's much less code compared to EKS, for example. You'll see the `main.tf` configuration first and then you'll take a look at `variables.tf`. Let's break down the following code:

1. First, you have the Google Terraform provider, for which you'll need to specify the GCP project ID and the region in which you want to deploy the GKE cluster:

    ```
    provider "google" {
      project     = var.project_id
      region      = var.region
    }
    ```

2. Next, you'll create the `google_container_cluster` resource, which is the GKE cluster. It'll specify the cluster name, region, and worker node count:

    ```
    resource "google_container_cluster" "primary" {
      name      = var.cluster_name
    ```

```
    location = var.region

    remove_default_node_pool = true
    initial_node_count       = 1

    network    = var.vpc_name
    subnetwork = var.subnet_name
}
```

3. The final resource to create is the `google_container_node_pool` resource, which is for creating the Kubernetes worker nodes. Here is where you can specify:

    - The worker node count:

```
resource "google_container_node_pool" "nodes" {
    name       = "${google_container_cluster.primary.name}-node-pool"
    location   = var.region
    cluster    = google_container_cluster.primary.name
    node_count = var.node_count
```

    - The GCP scopes (or services) that you want GKE to have access to:

```
    node_config {
      oauth_scopes = [
        "https://www.googleapis.com/auth/logging.write",
        "https://www.googleapis.com/auth/monitoring",
      ]

      labels = {
        env = var.project_id
      }
```

    - The VM type or size:

```
      machine_type = "n1-standard-1"
      tags         = ["gke-node", "${var.project_id}-gke"]
      metadata = {
        disable-legacy-endpoints = "true"
      }
```

```
  }
}
```

Putting it all together, you'll have a `main.tf` configuration that you can use to set up a GKE cluster.

4. Next, let's take a look at `variables.tf`, which will contain the following:

    *   The GCP project ID:

    ```
    variable "project_id" {
      type = string
      default = "gold-mode-297211"
    }
    ```

    *   The GCP region:

    ```
    variable "region" {
      type = string
      default = "us-east1"
    }
    ```

    *   The GCP VPC name that GKE will exist in:

    ```
    variable "vpc_name" {
      type = string
      default = "default"
    }
    ```

    *   The subnet name inside of the VPC that you want GKE to be attached to:

    ```
    variable "subnet_name" {
      type = string
      default = "default"
    }
    ```

    *   The code count (Kubernetes worker nodes):

    ```
    variable "node_count" {
      type = string
      default = 2
    }
    ```

- The GKE cluster name:

```
variable "cluster_name" {
  type = string
  default = "gkek8senv"
}
```

You're now ready to put the proceeding code into the appropriate `main.tf` and `variables.tf` configuration files to create your GKE environment.

GKE Autopilot

To wrap up the *GKE* section, let's quickly talk about GKE Autopilot. Autopilot is the same concept as EKS Fargate. It's serverless Kubernetes, which means you don't have to worry about managing the worker nodes for your GKE cluster. Instead, you only have to worry about deploying the application(s) and setting up any monitoring, logging, traces, alerts, and metrics you'd like to capture from the GKE cluster.

A quick note on multi-cloud

Many engineers just getting started with Kubernetes may not come across it too much, but multi-cloud is very much a reality. Just as organizations didn't want to rely on one data center for redundancy, some organizations don't want only one cloud for redundancy. Instead, they want to think about the multi-cloud approach – for example, scaling out Kubernetes workloads from AKS to GKE.

This implementation can be rather advanced and require a ton of security-related permissions, authentication and authorization capabilities between clouds, and heavy networking knowledge to ensure Kubernetes clusters between clouds can communicate with each other. Because of that, it's highly recommended to do extensive research before implementing this and ensure that all of the proper testing went as expected.

Summary

Although a multi-cloud approach may not be at the forefront of everyone's mind, it's still super crucial to understand how the three clouds work with Kubernetes. The reason why is that chances are, throughout your Kubernetes journey, you'll work in one cloud, but when the need arises to work in other clouds, you should be prepared.

In this chapter, you learned about setting up, managing, and maintaining Kubernetes clusters across Azure, AWS, and GCP. One of the biggest takeaways is that at the end of the day, the setup of Kubernetes across the clouds isn't really so different. They're all sort of doing the same thing with different service names.

Further reading

- *Building Google Cloud Platform Solutions* by Ted Hunter, Steven Porter, and Legorie Rajan PS:

  `https://www.packtpub.com/product/building-google-cloud-platform-solutions/9781838647438`

- *Hands-On Kubernetes on Azure – Second Edition* by Nills Franssens, Shivakumar Gopalakrishnan, and Gunther Lenz:

  `https://www.packtpub.com/product/hands-on-kubernetes-on-azure-second-edition/9781800209671`

- *Learning AWS – Second Edition* by Aurobindo Sarkar and Amit Shah:

  `https://www.packtpub.com/product/learning-aws-second-edition/9781787281066`

3

Running Kubernetes with Other Cloud Pals

Chances are that throughout this book, and perhaps even so far, you're going to get whiplash by finding out about the number of places and different ways you can deploy Kubernetes. The reality is, you're going to get even more whiplash in the real world. Whether you're a full-time Kubernetes engineer or a consultant, every company that you go to is going to feel a bit different in the ways you're deploying Kubernetes and where you're deploying it.

In the previous chapter, you learned about the three major Kubernetes cloud services – AKS, EKS, and GKE. However, there are a ton of other great options in the wild that are on private clouds and **Platform-as-a-Service (PaaS)** solutions. Although you'll see a lot of organizations, ranging from start-ups to Fortune 200 companies and up, using popular Kubernetes cloud-based services such as AKS, EKS, and GKE, more and more organizations are starting to use private clouds for secondary Kubernetes clusters or even to save money because the larger cloud providers are typically far more expensive. Taking it to another level, some organizations are completely ditching the idea of having a Kubernetes cluster in the cloud and going PaaS, and you'll see why in the upcoming sections.

By the end of this chapter and with the help of the previous chapter, you'll be able to identify what solution your organization should go with and why it would be useful. From an individual perspective, you'll walk away from this chapter with the know-how of multiple managed Kubernetes offerings. That way, you'll be far more marketable in the job space and use all the different platforms.

In this chapter, we're going to cover the following topics:

- Understanding Linode Kubernetes Engine
- Exploring DigitalOcean Managed Kubernetes
- What is Kubernetes PaaS and how does it differ?

Technical requirements

For this chapter, you should already know a bit about cloud technologies. The gist is that all clouds are more or less the same. There are differences in the names of the services, but they're all doing the same thing more or less.

If you're comfortable with the cloud and have worked in a few cloud-based services, you'll be successful in navigating this chapter.

To work inside the cloud-based services, you will need the following:

- A Linode account
- A DigitalOcean account
- A Red Hat account
- An AWS account (for the final section of this chapter)

You can sign up for all of these services and get free credit. Just ensure that you shut down the Kubernetes environments when you've finished running them to save money.

The code for this chapter can be found in this book's GitHub repository at `https://github.com/PacktPublishing/50-Kubernetes-Concepts-Every-DevOps-Engineer-Should-Know/tree/main/Ch3`.

Understanding Linode Kubernetes Engine

Linode, recently acquired by Akami Technologies, is a developer-friendly private cloud that is very well-known for its easy dashboard and feature-rich platform that isn't overly complex. Linode focuses on ease of use with a cloud -for -all mindset. Some key callouts for Linode include transparent pricing with almost zero guesswork, easily scalable workloads, a full/public API, and a GUI-based cloud manager. Most of all, Linode is known for its *always human* customer support.

When it comes to comparing Linode and other private cloud providers, Linode sticks out by offering cloud GPUs and high outbound transfer speeds, along with its customer support.

In this section, you're going to learn about why you'd want to use **Linode Kubernetes Engine** (LKE) and how to set up the LKE portal, create a Kubernetes cluster in LKE manually, take the same manual process to automate it, and deploy your Kubernetes workloads.

Why LKE?

When you're choosing a cloud, the last thing you want is to have to guess how much your monthly bill is going to be. This is why people are nervous about the cloud and even more nervous about serverless technologies. The monthly cost can be unknown, which isn't the best answer to give to a CFO. With Linode, costs are bundled together, so you know exactly what you're going to pay for, and that's very important for billing administrators and engineers alike. When it comes to scaling, both horizontally and vertically, the last thing that any engineer wants to have to sit and figure out manually is how much the environment is going to cost the company every month.

Another large cost saving is with the Control Plane. Much like any Kubernetes cloud service, the idea is to abstract the Kubernetes Control Plane/API server away from you. That way, you don't have to worry about managing anything other than worker nodes and the application(s). Linode doesn't charge for the Control Plane, whereas other clouds do. For example, EKS and GKE charge a per cluster management fee of $10 per hour or $73.00 per month. Although this may not seem like a lot, for a start-up that's getting by with bootstrap funding and has enough bills, they most likely don't want one more.

Setting up LKE manually

Now that you know the theory behind why you'd want to choose Linode, along with some pricing metrics and other aspects that make Linode great, it's time to get hands-on and learn about setting up LKE.

For this section, ensure that you are signed into Linode via a web browser of your choosing. Follow these steps:

1. On the Linode dashboard, choose **Kubernetes**:

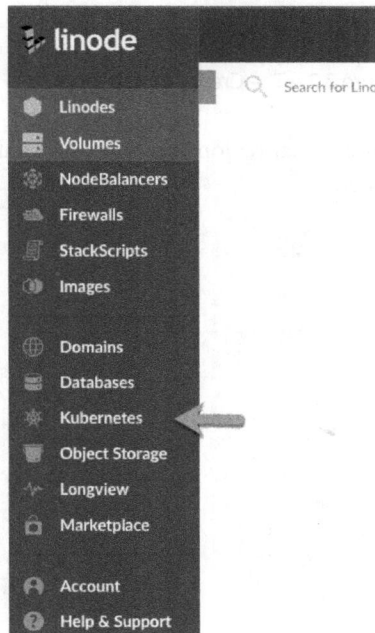

Figure 3.1 – The LKE portal

2.   Click the blue **Create Cluster** button:

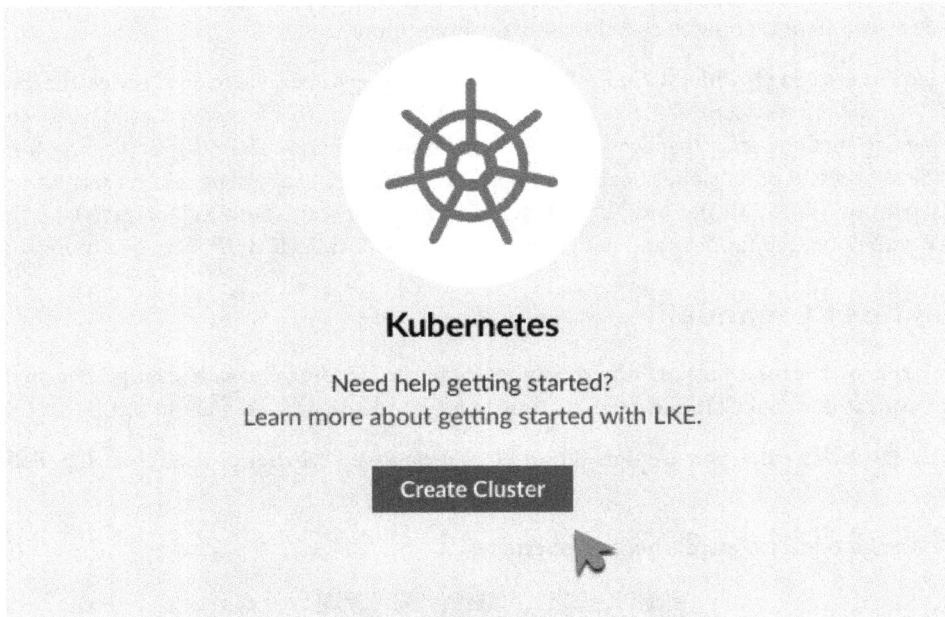

Figure 3.2 – The Create Cluster button

3.   Choose a name for your cluster, what region/location you want the LKE cluster to reside in, and the Kubernetes API version:

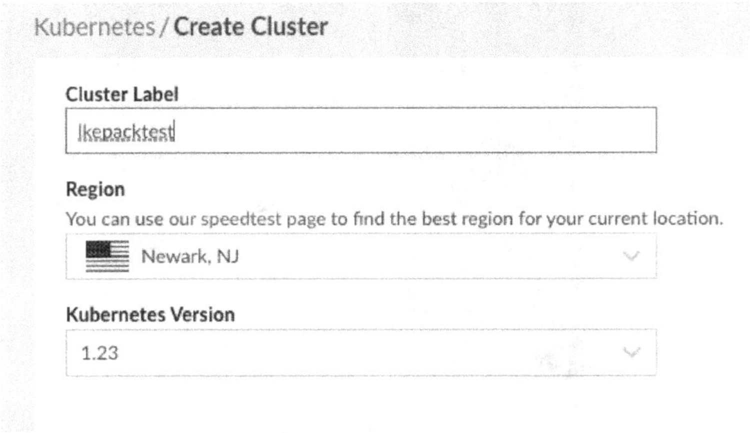

Figure 3.3 – Adding a Cluster label, region, and Kubernetes version

4. When choosing node pools, you have a few options:

    - **Dedicated CPU**: Good for workloads where consistent performance is crucial for daily workflows
    - **Shared CPU**: Good for medium workflows, such as a secret engine (something that isn't getting a lot of traffic)
    - **High Memory**: Good for RAM-intensive applications, such as older Java applications, in-memory databases, and cached data

    For this section, you can choose **Shared CPU** as it's the most cost-effective:

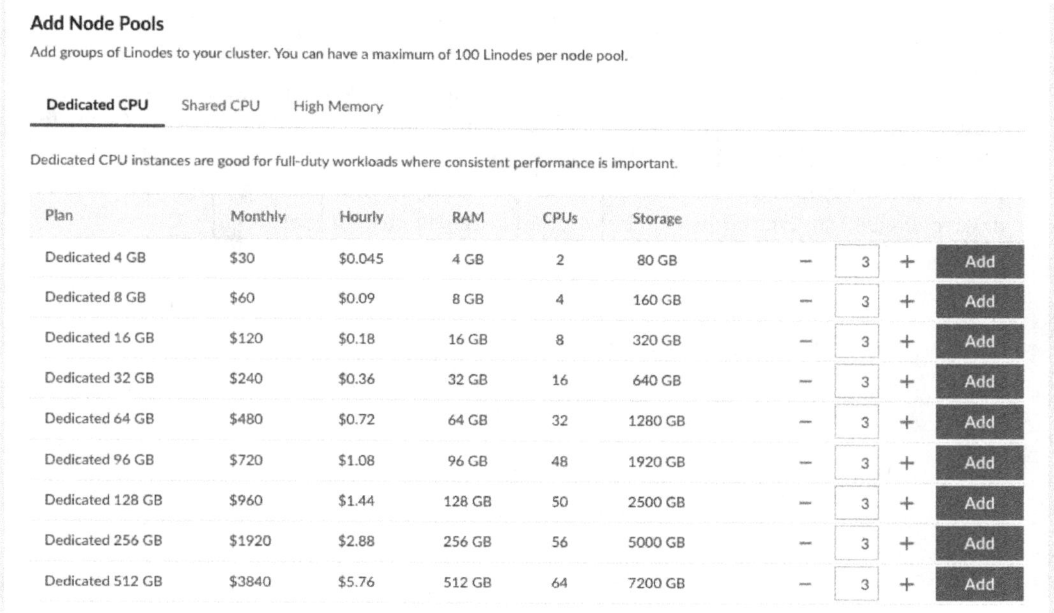

Add Node Pools

Add groups of Linodes to your cluster. You can have a maximum of 100 Linodes per node pool.

Dedicated CPU   Shared CPU   High Memory

Dedicated CPU instances are good for full-duty workloads where consistent performance is important.

Plan	Monthly	Hourly	RAM	CPUs	Storage				
Dedicated 4 GB	$30	$0.045	4 GB	2	80 GB	—	3	+	Add
Dedicated 8 GB	$60	$0.09	8 GB	4	160 GB	—	3	+	Add
Dedicated 16 GB	$120	$0.18	16 GB	8	320 GB	—	3	+	Add
Dedicated 32 GB	$240	$0.36	32 GB	16	640 GB	—	3	+	Add
Dedicated 64 GB	$480	$0.72	64 GB	32	1280 GB	—	3	+	Add
Dedicated 96 GB	$720	$1.08	96 GB	48	1920 GB	—	3	+	Add
Dedicated 128 GB	$960	$1.44	128 GB	50	2500 GB	—	3	+	Add
Dedicated 256 GB	$1920	$2.88	256 GB	56	5000 GB	—	3	+	Add
Dedicated 512 GB	$3840	$5.76	512 GB	64	7200 GB	—	3	+	Add

Figure 3.4 – Worker node size

5. To continue to keep things cost-effective, choose the **Linode 2 GB** option and ensure you scale it down to *1* node:

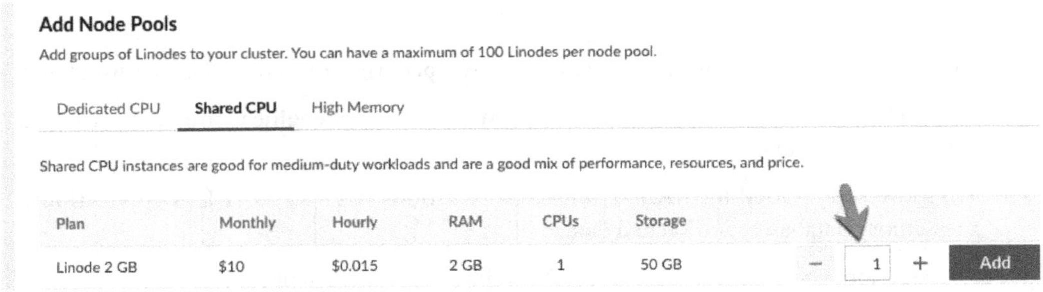

Add Node Pools

Add groups of Linodes to your cluster. You can have a maximum of 100 Linodes per node pool.

Dedicated CPU   **Shared CPU**   High Memory

Shared CPU instances are good for medium-duty workloads and are a good mix of performance, resources, and price.

Plan	Monthly	Hourly	RAM	CPUs	Storage				
Linode 2 GB	$10	$0.015	2 GB	1	50 GB	−	1	+	Add

Figure 3.5 – The Add Node Pools page

6. In any production-level environment, you always want to think about **high availability** (**HA**). When it comes to Kubernetes, it's no different. LKE offers the ability to enable HA for the Kubernetes Control Plane. For production environments, you'll 100% want to implement this. For lab/dev environments (which is what you're building now for learning purposes), you don't have to enable HA. Once done, click the blue **Create Cluster** button:

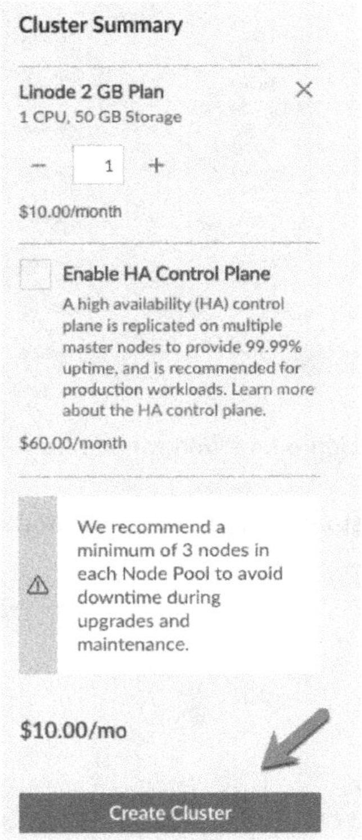

Cluster Summary

Linode 2 GB Plan   ✕
1 CPU, 50 GB Storage

−   1   +

$10.00/month

☐ **Enable HA Control Plane**

A high availability (HA) control plane is replicated on multiple master nodes to provide 99.99% uptime, and is recommended for production workloads. Learn more about the HA control plane.

$60.00/month

⚠ We recommend a minimum of 3 nodes in each Node Pool to avoid downtime during upgrades and maintenance.

$10.00/mo

Create Cluster

Figure 3.6 – HA Control Plane

Now that you're familiar with the manual process of creating an LKE cluster, it's time to learn how to automate it and make the process repeatable for production-level environments.

Automating LKE deployments

Now that you know how to create an LKE cluster manually, it's time to learn how to create it with Terraform so you can ensure repeatable processes throughout your environment. In many production-level cases, you'll run the following Terraform code within a CI/CD pipeline to ensure repeatability. For this section, you can run it locally.

First, you'll see the `main.tf` configuration and then look at `variables.tf`.

First, there's the Terraform provider. The provider will utilize the newest version of the Linode Terraform provider. For Terraform to interact with the Linode API, you'll need to pass in an API key that you can create from your Linode account:

```
terraform {
  required_providers {
    linode = {
      source = "linode/linode"
    }
  }
}


provider "linode" {
  token = var.token
}
```

Next, there's the `linode_lke_cluster` resource, which will create the LKE cluster. Within the dynamic block, you'll see a `for_each` loop that specifies how many worker nodes will be created based on the pool amount. The pool amount is the number of worker nodes you want to deploy (between 3 to 4 is recommended for production):

```
resource "linode_lke_cluster" "packtlke" {
    k8s_version = var.apiversion
    region = var.region

    dynamic "pool" {
        for_each = var.pools
        content {
```

```
            type  = pool.value["type"]
            count = pool.value["count"]
        }
    }
}
```

The last piece of code is the output of `kubeconfig`, which contains all of the authentication and authorization configurations to connect to the Kubernetes cluster:

```
output "kubeconfig" {
    value = linode_lke_cluster.packtlke.kubeconfig
    sensitive = true
}
```

Now that you have the main Terraform configuration, you'll need variables to pass in. These variables allow your code to stay repeatable so that you don't have to continuously change hardcoded values or create new configurations for each environment. The reason why is that due to formatting, it may look out of the ordinary on your page while you're reading this chapter.

For this section, you can take a look at the variables on GitHub at `https://github.com/PacktPublishing/50-Kubernetes-Concepts-Every-DevOps-Engineer-Should-Know/blob/main/Ch3/LKE/variables.tf`.

Although these are all standard Terraform variables and don't require much explanation, the one variable to point out is the `pools` variable. Notice how there's a list type specified for the variable, which includes how many worker nodes and the size of the worker nodes on Linode. The reason why the variable is a list type is that in the `main.tf` configuration, the `dynamic "pool"` block calls for a list when using the `for` loop.

One thing to keep in mind when it comes to LKE is understanding Linode. Although Linode is a great cloud provider, the truth is, it's not going to have as many services and features for Kubernetes as, for example, EKS. Taking EKS as an example, there are IAM roles and RBAC-related permissions you can configure, DNS management with Route53, Secrets management, a container registry, and Fargate profiles for serverless Kubernetes. Even Azure and GCP have very similar services. With a provider such as Linode, however, they don't. That's not to discount Linode or say that they aren't a good Kubernetes provider because the truth is, they very much are. However, a situation such as Linode not having IAM/RBAC built-in capabilities may be a deal breaker for many production engineering and security teams.

Now that you know how to create an LKE cluster both manually and automatically, it's time to move on to the next section.

Exploring DigitalOcean Managed Kubernetes

DigitalOcean, much like Linode, markets toward the notion of an easy cloud to use compared to other large clouds with (what feels like) millions of services to choose from. DigitalOcean's slogan is *Simpler cloud. Happier devs. Better results.* Over the years, DigitalOcean wasn't only known for its cloud platform, but its blogs and how-to guides. DigitalOcean, for many engineers, became the standard go-to online location for learning how to do something in a hands-on fashion. Many writers use the DigitalOcean Technical Writing Guidelines that DigitalOcean created for writer/blogger best practices.

In this section, you're going to learn about why you'd want to use **DigitalOcean Managed Kubernetes**, the pros of the Kubernetes service, setting up DigitalOcean Managed Kubernetes manually, and taking the same manual process, but doing it in an automated fashion with Terraform.

Why DigitalOcean Kubernetes Engine?

Since DigitalOcean was founded in 2011, developers around the globe have been using it for its ease of use and straightforward deployments. A lot of engineers even use DigitalOcean for hosting their projects (personal websites, blogs, servers, and so on). It's far easier in many cases than having to worry about creating a bunch of services in a large public cloud.

From an ease-of-use perspective, DigitalOcean Kubernetes Engine does not disappoint. Much like any other Kubernetes service, the purpose is to abstract away the need to manage the underlying Control Plane/API server. The whole idea here is to lower the barrier of entry when it comes to using a Kubernetes service.

Compared to other products such as EKS/GKE/AKS, DigitalOcean Kubernetes Engine is more focused on the Day Two operations piece of Kubernetes. The complexities of a lot of the Kubernetes services out in the wild sometimes make engineers run away because they want something that *just works* out of the box.

> **Important note**
> DigitalOcean Managed Kubernetes, although easy to use, appears to be a bit out of date from a Kubernetes API perspective compared to its counterparts. Whereas many Kubernetes services offer Kubernetes API version v1.23 and above, DigitalOcean only offers up to v1.22.8 at the time of writing. Keep this in mind and remember to check as you may need different API versions.

Setting up DigitalOcean Managed Kubernetes manually

Now that you know the theory behind why you'd want to choose DigitalOcean, along with some pricing metrics and other aspects that make DigitalOcean great, it's time to get hands-on and learn about setting up DigitalOcean Kubernetes Engine.

For this section, ensure that you are signed into DigitalOcean via a web browser of your choosing. Follow these steps:

1. On the DigitalOcean dashboard, choose **Kubernetes**:

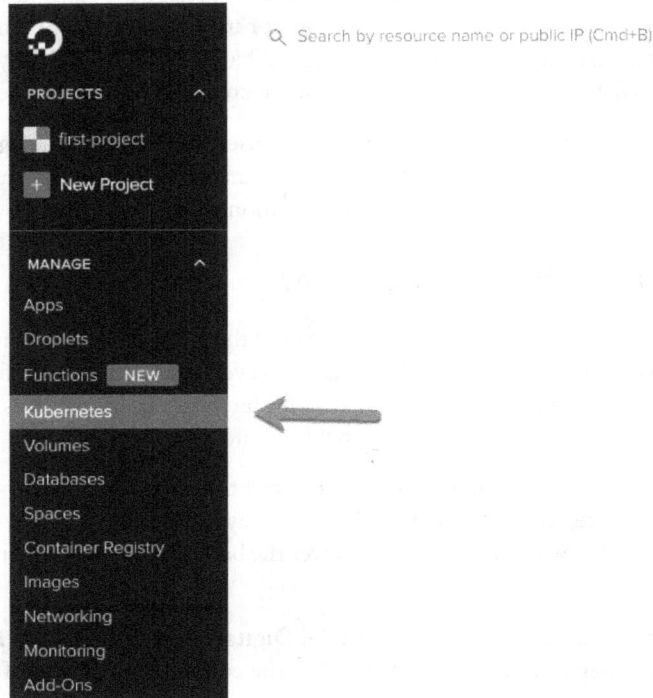

Figure 3.7 – DigitalOcean Managed Kubernetes

2. Click the blue **Create a Kubernetes Cluster** button:

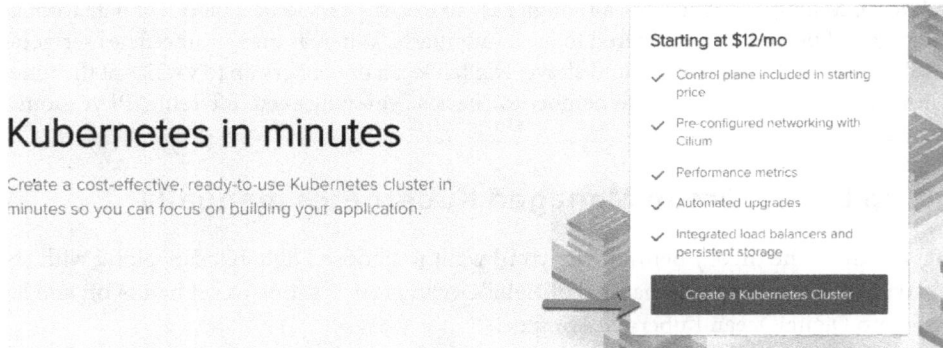

Figure 3.8 – The Create a Kubernetes Cluster button

3. Choose your region, VPC name, and Kubernetes API version. The recommended API version, in general, is to always go with the latest unless you have a specific reason not to (the same rule applies to any Kubernetes environment):

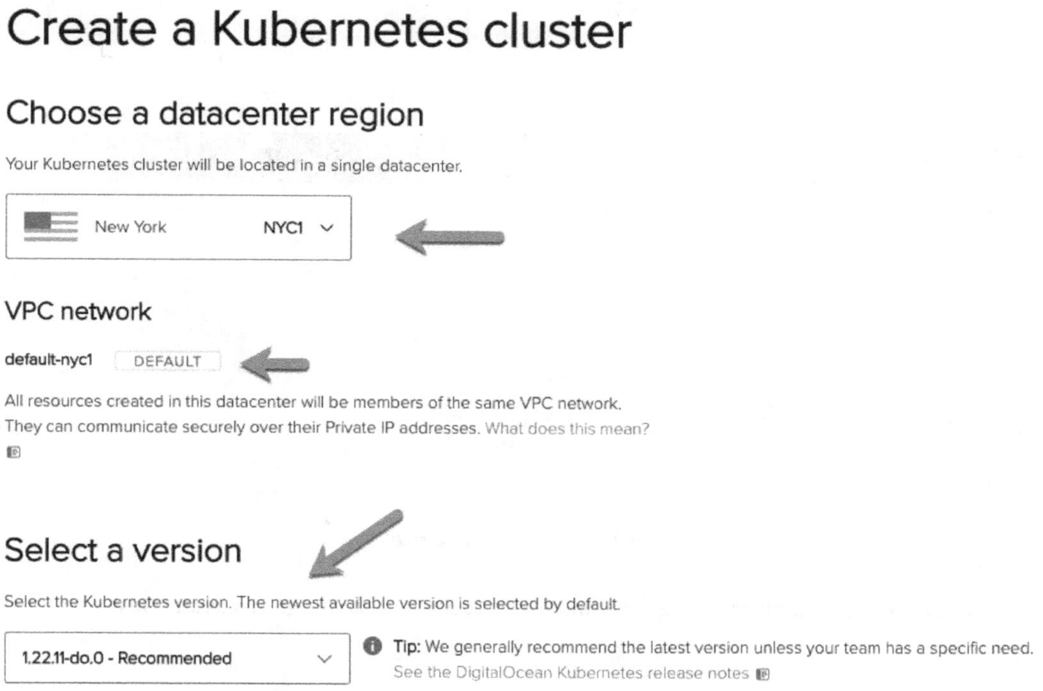

Create a Kubernetes cluster

Choose a datacenter region

Your Kubernetes cluster will be located in a single datacenter.

VPC network

default-nyc1   DEFAULT

All resources created in this datacenter will be members of the same VPC network. They can communicate securely over their Private IP addresses. What does this mean?

Select a version

Select the Kubernetes version. The newest available version is selected by default.

1.22.11-do.0 - Recommended   ⌄

ⓘ **Tip:** We generally recommend the latest version unless your team has a specific need. See the DigitalOcean Kubernetes release notes

Figure 3.9 – Adding cluster details

4. Choose the cluster capacity. The two very important sections here are as follows:

   *   **Machine type:** For this, you'll have to choose what's best for you and your production environment. Although DigitalOcean doesn't have as many options as Linode, you can choose from a basic node, Intel-based node, or AMD-based node from a CPU perspective.

   *   **High availability Control Plane:** For this, you'll always want to ensure that the Control Plane is highly available. The Control Plane holds the scheduler, etcd, and many other important Kubernetes components. Without them, Kubernetes wouldn't work.

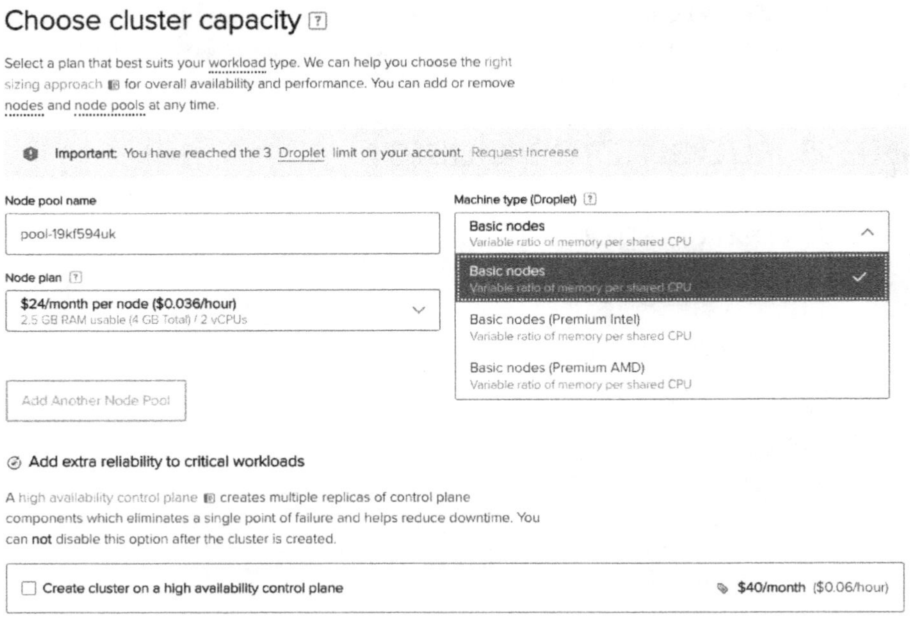

Choose cluster capacity ⏣

Select a plan that best suits your workload type. We can help you choose the right sizing approach ⏣ for overall availability and performance. You can add or remove nodes and node pools at any time.

> ⬢ **Important:** You have reached the 3 Droplet limit on your account. Request Increase

Node pool name	Machine type (Droplet) ⏣
pool-19kf594uk	**Basic nodes** — Variable ratio of memory per shared CPU ∧

Node plan ⏣

$24/month per node ($0.036/hour)   ∨
2.5 GB RAM usable (4 GB Total) / 2 vCPUs

Basic nodes — Variable ratio of memory per shared CPU   ✓

Basic nodes (Premium Intel) — Variable ratio of memory per shared CPU

Basic nodes (Premium AMD) — Variable ratio of memory per shared CPU

Add Another Node Pool

⊘ **Add extra reliability to critical workloads**

A high availability control plane ⏣ creates multiple replicas of control plane components which eliminates a single point of failure and helps reduce downtime. You can **not** disable this option after the cluster is created.

☐ Create cluster on a high availability control plane   🏷 **$40/month** ($0.06/hour)

Figure 3.10 – Worker node size

5.   Confirm your cluster by reviewing the monthly charge and clicking the green **Create Cluster** button:

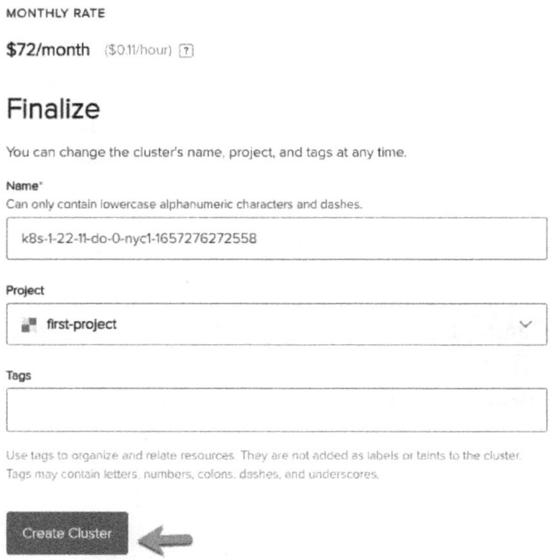

MONTHLY RATE

$72/month   ($0.11/hour) ⏣

Finalize

You can change the cluster's name, project, and tags at any time.

Name*
Can only contain lowercase alphanumeric characters and dashes.

k8s-1-22-11-do-0-nyc1-1657276272558

Project

▞ first-project   ∨

Tags

Use tags to organize and relate resources. They are not added as labels or taints to the cluster. Tags may contain letters, numbers, colons, dashes, and underscores.

Create Cluster ⟵

Figure 3.11 – Finalizing the cluster

Now that you're familiar with the manual process of creating a DigitalOcean Kubernetes Engine cluster, it's time to learn how to automate it and make the process repeatable for production-level environments.

Automating DigitalOcean Managed Kubernetes

From an automation perspective, you have a few óptions. Two of the most popular are the DigitalOcean CLI and **Infrastructure as Code (IaC)**. In this section, you'll learn how to create a DigitalOcean Managed Kubernetes cluster using Terraform.

In many production-level cases, you'll run the following Terraform code within a CI/CD pipeline to ensure repeatability. For this section, you can run it locally.

Like we did for LKE, first, we'll see the `main.tf` configuration and then you'll take a look at `variables.tf`.

The Terraform configuration starts as all others do: with the provider. The DigitalOcean Terraform provider requires you to pass in a DigitalOcean API token, which you can generate from the DigitalOcean UI:

```
terraform {
  required_providers {
    digitalocean = {
      source = "digitalocean/digitalocean"
    }
  }
}


provider "digitalocean" {
  token = var.do_token
}
```

Next, one resource block is needed, which is used to create the entire cluster and the node pools. These are DigitalOcean Droplets that end up being Kubernetes worker nodes. It also creates horizontal auto-scaling. For some DigitalOcean accounts, the maximum Droplet amount is three, so you'll most likely want to increase that for production environments:

```
resource "digitalocean_kubernetes_cluster" "packtdo" {
  name    = var.cluster_name
  region  = var.region
  version = var.k8s_version

  node_pool {
```

```
    name       = "autoscale-worker-pool"
    size       = "s-2vcpu-2gb"
    auto_scale = true
    min_nodes  = 2
    max_nodes  = 3
  }
}
```

Now that you have the Terraform configuration, you'll need variables to pass in. These variables allow your code to stay repeatable so that you don't have to continuously change hardcoded values or create new configurations for each environment.

There are four variables, as follows:

- `region`: The region that the DigitalOcean Kubernetes Engine cluster will run in.
- `cluster_name`: The name of the Kubernetes cluster.
- `K8s_version`: The Kubernetes API version.
- `do_token`: The DigitalOcean API token. For production-level environments, you'll want to store this in some type of secret store and have Terraform retrieve it with a data block. Writing an API token into a variable and pushing it up to source control is a big no-no:

```
variable "region" {
    type = string
    default = "nyc1"
}

variable "cluster_name" {
    type = string
    default = "packtdo01"
}

variable "k8s_version" {
    type = string
    default = "1.22.11-do.0"
}

variable "do_token" {
    type = string
```

```
    default = ""
    sensitive = true
}
```

Wrapping up this section on DigitalOcean, one thing to keep in mind is the same piece that was said in the *Understanding Linode Kubernetes Engine* section – bigger cloud providers are going to have more services that can tie into the managed Kubernetes offerings. This is something you'll have to keep in mind as you decide what's going to work best for your environment.

In the next and final section of this chapter, you'll learn about PaaS with OpenShift from a theoretical and hands-on perspective.

What is Kubernetes PaaS and how does it differ?

Deploying Kubernetes clusters in different ways felt to engineers like it came in waves. First, there were raw Kubernetes clusters. You'd have to deploy everything manually, ranging from the Control Plane to the **Certificate Authority** (**CA**) and everything in between. After that, there were Kubernetes services in the cloud, such as AKS, GKE, and EKS. Now, there are serverless Kubernetes such as GKE AutoPilot and EKS Fargate, which you learned about in the previous chapter.

Another option that stands out, especially in the enterprise, is PaaS-based Kubernetes solutions such as Red Hat's **OpenShift**.

In this section, you're going to learn about why you'd want to use OpenShift, how enterprises are utilizing PaaS-based Kubernetes such as OpenShift, and how to get started with a Dev environment right on your local computer with OpenShift, develop and deploy production-ready OpenShift clusters in major cloud providers, and deploy production-ready applications inside of OpenShift.

OpenShift

OpenShift is an odd paradox between full-blown Kubernetes and its own orchestration system. Underneath the hood, OpenShift uses Kubernetes. If you write a Kubernetes manifest for a Deployment, Pod, and so on, you can use it on OpenShift. Fundamentally, nothing changes when it comes to Kubernetes and OpenShift. However, there are differences in how you manage OpenShift versus how you manage Kubernetes. OpenShift is a PaaS, whereas Kubernetes can be managed with a cloud provider, so it sort of feels like **Software as a Service** (**SaaS**) and can be managed from a bare-metal perspective. Because Kubernetes is such a versatile platform, it can't be put into one category.

One thing you must remember when it comes to OpenShift is that it's enterprise-specific. There's no reason that an engineer would run OpenShift for a lab environment other than to learn (which is what you're doing in this chapter). With Kubernetes, you have far more deployment options and options regarding where you can deploy it. With OpenShift, you're limited to a certain type of virtual machine and where/how you can deploy it. This isn't necessarily a bad thing in the slightest. OpenShift wasn't meant for engineers to do labs with like minikube and Docker Desktop. It was built with enterprise customers in mind. If you're interested in diving deeper into this topic, I highly recommend reading this blog post from Tomasz Cholewa on comparing Kubernetes with OpenShift: `https://blog.cloudowski.com/articles/10-differences-between-openshift-and-kubernetes/`.

The definition of OpenShift, as per Red Hat, is that *"Red Hat OpenShift delivers a complete application platform for both traditional and cloud-native applications, allowing them to run anywhere. Built on Red Hat Enterprise Linux and compatible with Red Hat Ansible Automation Platform, Red Hat OpenShift enables automation inside and outside your Kubernetes clusters."*

Simply put, it allows you to orchestrate and manage containerized applications in a PaaS environment.

OpenShift in the enterprise

At this point, you may be wondering why anyone would want to use OpenShift over a standard Kubernetes deployment. Kubernetes has a ton of support, is supported by all major cloud providers, and is the latest and greatest. When it comes to the enterprise, Kubernetes is thought of a bit differently.

To leadership teams, Kubernetes is often thought of as a black box of magic and mystery that's going to cost them a ton of money to maintain and support. The reality is that in enterprise environments, leadership teams want the ability to call a support number or contact an account executive when something breaks. They want the *enterprise software* so that if (when) something goes wrong, they know that the engineering teams have someone to call. Even though engineers will most likely spend more time waiting to hear back from support than doing it themselves, *enterprise licensing* gives leadership teams peace of mind. However, with peace of mind comes cost. OpenShift licensing is very expensive and remember, you have to run it somewhere, which will cost you money as well. If you run OpenShift in, for example, AWS, you're paying for the cloud infrastructure running in AWS, OpenShift licensing, and Red Hat support. If you decide to go the OpenShift route, ensure that your leadership teams understand the cost.

From a technical and engineering perspective, OpenShift isn't doing anything differently than what Kubernetes can do. Sure, to have Kubernetes do exactly what OpenShift does would require some work and engineering efforts to build it, but it's all very much doable. Although OpenShift is a great platform, it's not doing anything overly extraordinary compared to Kubernetes.

Getting started with OpenShift Sandbox

Before spending money on OpenShift, you can test it out using OpenShift ReadyContainers in a sandbox environment. Although the sandbox environment is not production-ready, it's a great way to test out and familiarize yourself with how OpenShift works. It's also great for lab environments! Follow these steps:

> **Important note**
> If you're on an M1 Mac, OpenShift ReadyContainers are not currently supported for ARM devices, so this lab won't work for you.

1. Log into the Red Hat console: `https://console.redhat.com/`.

2. Click on **OpenShift**:

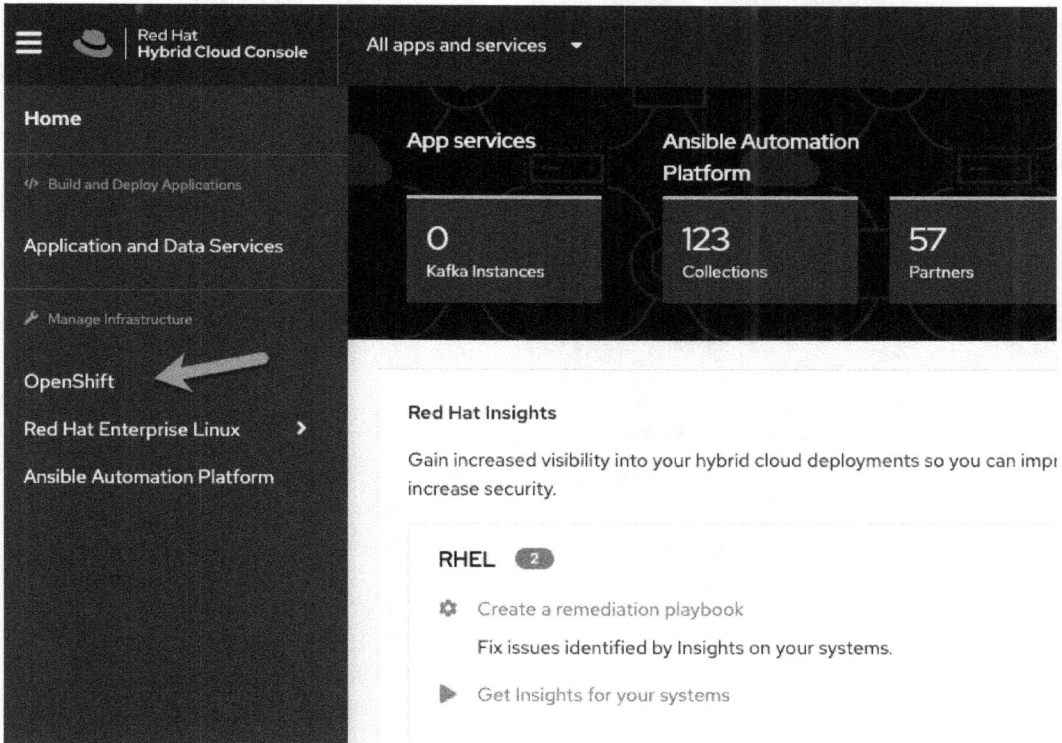

Figure 3.12 – Red Hat console

3.   Choose the **Clusters** option:

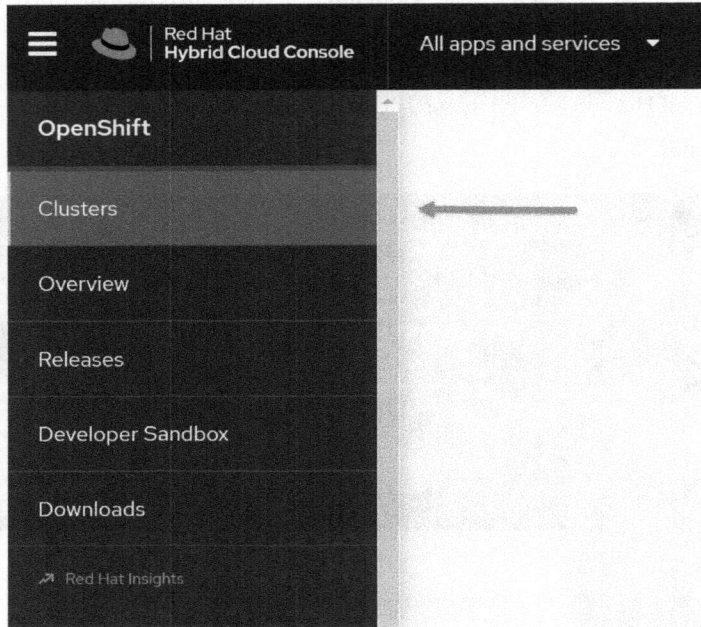

Figure 3.13 – Clusters

4.   Under the **Clusters** option, you'll see three options – **Cloud**, **Datacenter**, and **Local**. Choose **Local**:

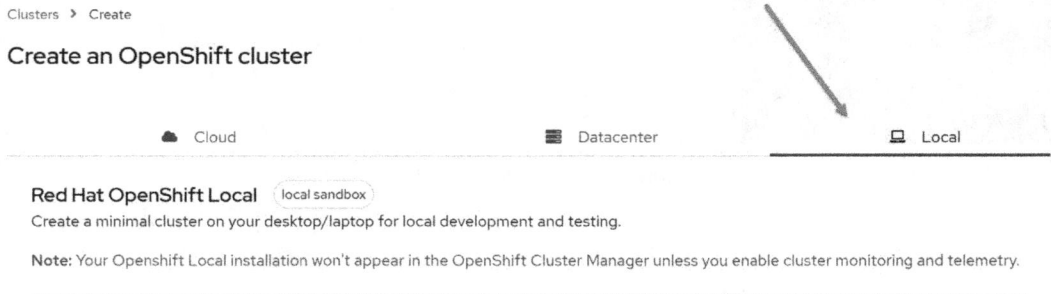

Figure 3.14 – Local cluster

5.   Download OpenShift locally by clicking the blue **Download OpenShift Local** button:

(1)   **Download what you need to get started**

OpenShift Local

Download and extract the OpenShift Local archive on your computer and open the installer. Opening the installer will automatically start a step-by-step installation guide.

| Windows ▼ | x86_64 ▼ | Download OpenShift Local |

Pull secret

Download or copy your pull secret. You'll be prompted for this information during installation.

| Download pull secret | 📋 Copy pull secret

Figure 3.15 – Download OpenShift Local

6.   Once OpenShift Local has been installed, you will need to run two commands (the instructions for installing CRC can be found at `https://crc.dev/crc/#minimum-system-requirements-operating-system_gsg`):

    A.   `crc setup`: Set up the configuration to authenticate to Red Hat

    B.   `crc start`: Start the local OpenShift cluster:

(2)   **Follow the documentation to install OpenShift Local**

Run `crc setup` to set up your host operating system for the OpenShift Local virtual machine.

Then, run `crc start` to create a minimal OpenShift 4 cluster on your computer.

View the OpenShift Local Getting started guide 🗗

Figure 3.16 – OpenShift Local setup

7.   Once you run the `crc start` command, you'll see an output on your terminal similar to the following:

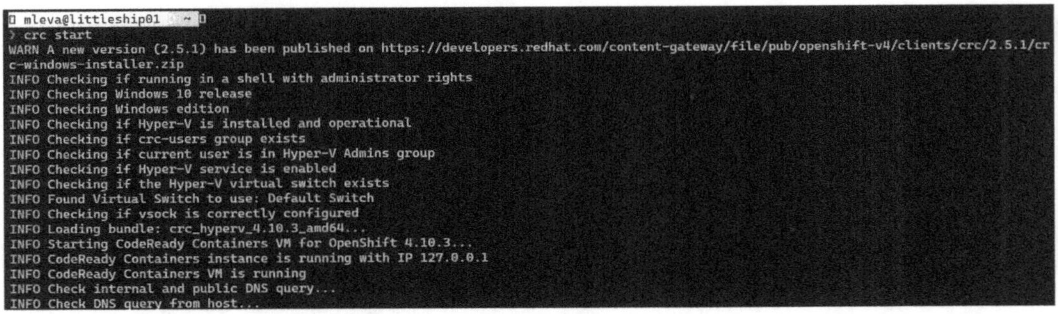

Figure 3.17 – Starting OpenShift Local

Depending on when you're reading this and based on version changes, you may need some configurations, including passing in an OpenShift token to authenticate from your localhost. To keep these steps brief and since this information is already available from Red Hat, you can follow the installation instructions here: `https://access.redhat.com/documentation/en-us/red_hat_openshift_local/2.5/html/getting_started_guide/installation_gsg#installing_gsg`.

Although we didn't touch on it in this section, there's a second lab environment option known as OpenShift Sandbox, which is different than ReadyContainers. You can set up OpenShift Sandbox here: `https://developers.redhat.com/developer-sandbox`.

Now that you know how to get an OpenShift sandbox up and running, let's learn how to set up a production-ready OpenShift cluster on AWS.

OpenShift on AWS

CodeReady containers are amazing because they allow you to utilize your local computer to learn OpenShift, much like Minikube allows you to learn Kubernetes and Docker Desktop allows you to learn Docker for free.

Now that you know about the free version, let's quickly dive into how to deploy OpenShift to the cloud. In this example, you'll learn about AWS, but the other cloud providers that are supported have the same workflow.

For this section, ensure that you are logged into your AWS console via the AWS CLI and that you are also logged into your Red Hat account. Follow these steps:

1.   Log into the Red Hat console: `https://console.redhat.com/`.

2.   Click on **OpenShift**:

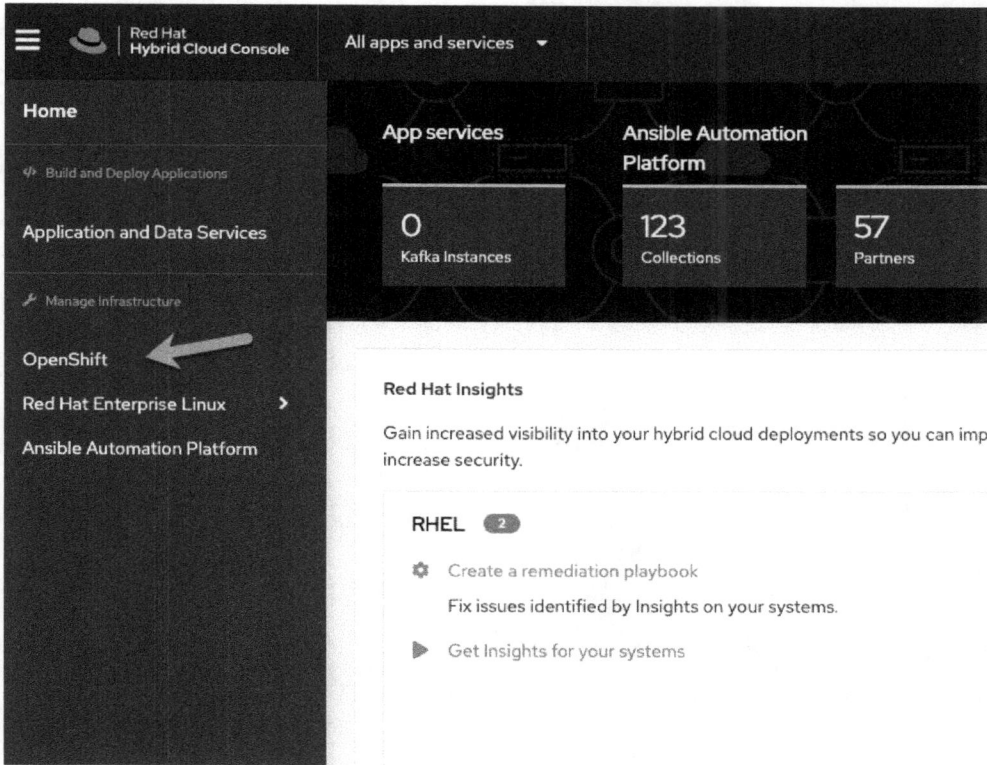

Figure 3.18 – OpenShift manager

3. Click the **Clusters** button and then click the blue **Create cluster** button:

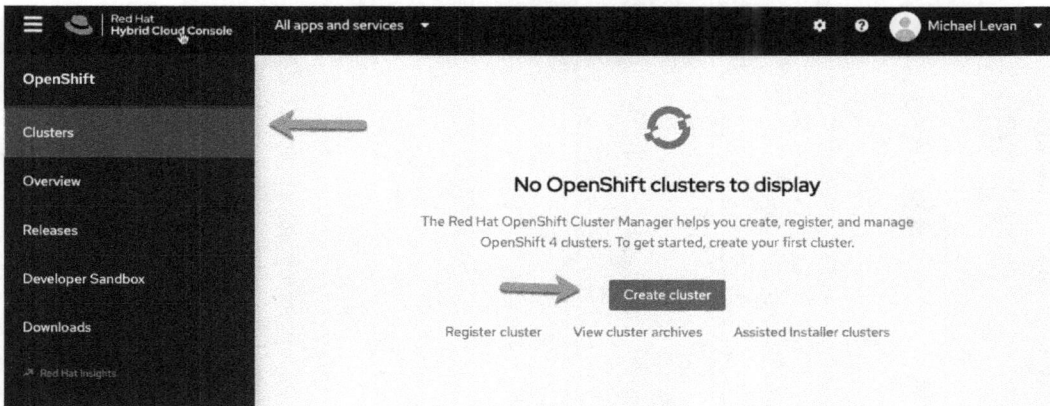

Figure 3.19 – Creating a cluster

4. You'll see several options to choose from, including Azure and IBM Cloud. Click the blue **Create cluster** button under the AWS option:

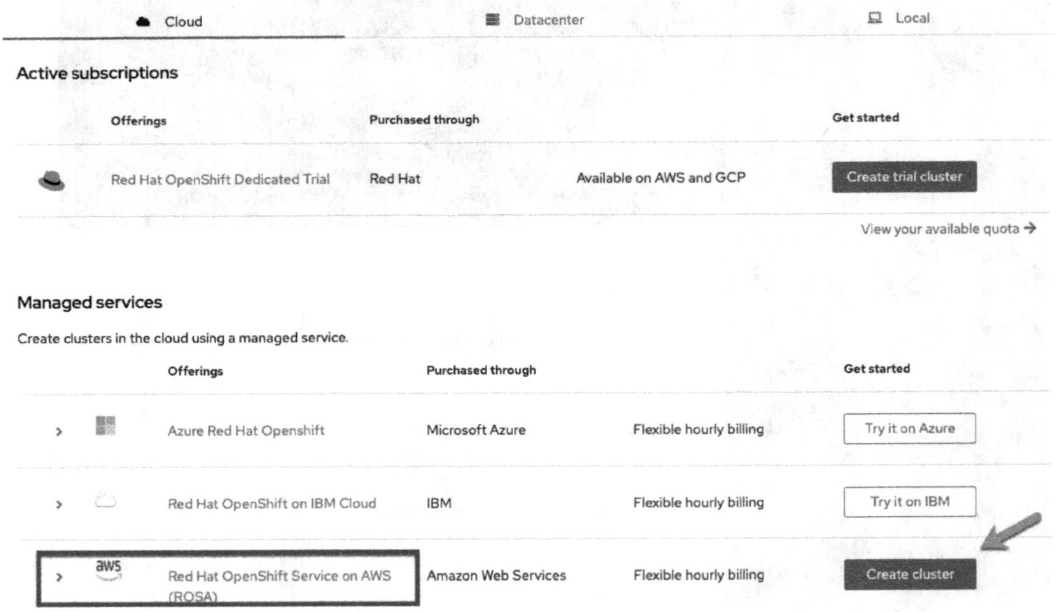

☁ Cloud	☰ Datacenter	🖥 Local

Active subscriptions

	Offerings	Purchased through		Get started
🎩	Red Hat OpenShift Dedicated Trial	Red Hat	Available on AWS and GCP	Create trial cluster

View your available quota →

Managed services

Create clusters in the cloud using a managed service.

		Offerings	Purchased through		Get started
>	▦	Azure Red Hat Openshift	Microsoft Azure	Flexible hourly billing	Try it on Azure
>	☁	Red Hat OpenShift on IBM Cloud	IBM	Flexible hourly billing	Try it on IBM
>	aws	Red Hat OpenShift Service on AWS (ROSA)	Amazon Web Services	Flexible hourly billing	Create cluster

Figure 3.20 – AWS ROSA

5. The first page you'll see associates your AWS account with Red Hat if you haven't done so already. To do that, click the **Select an account** button and go through the walk-through of configuring ROSA, which is the Red Hat OpenShift service on AWS:

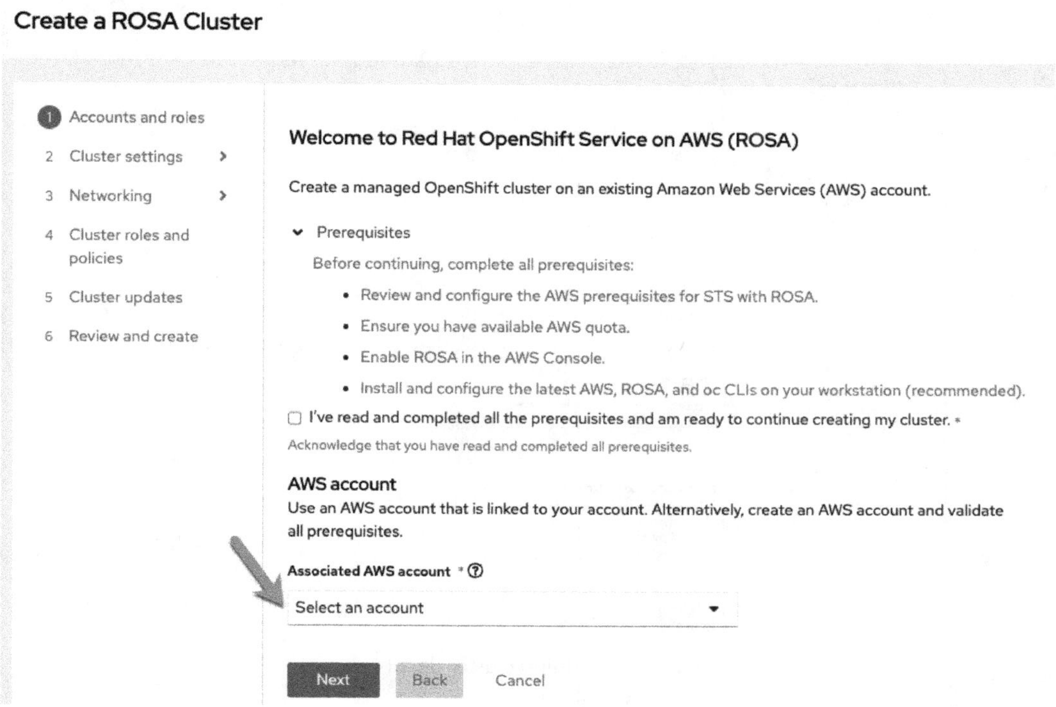

Create a ROSA Cluster

Figure 3.21 – The Create a ROSA Cluster page

6.   Once you've set up the AWS account permissions and roles for ROSA, the next page is all about configuring the cluster, which includes the cluster name, OpenShift version, region, and availability options. One of the cool options that OpenShift gives you is the ability to encrypt etcd and create persistent volumes with customer keys. This added security is typically looked at closely within the enterprise:

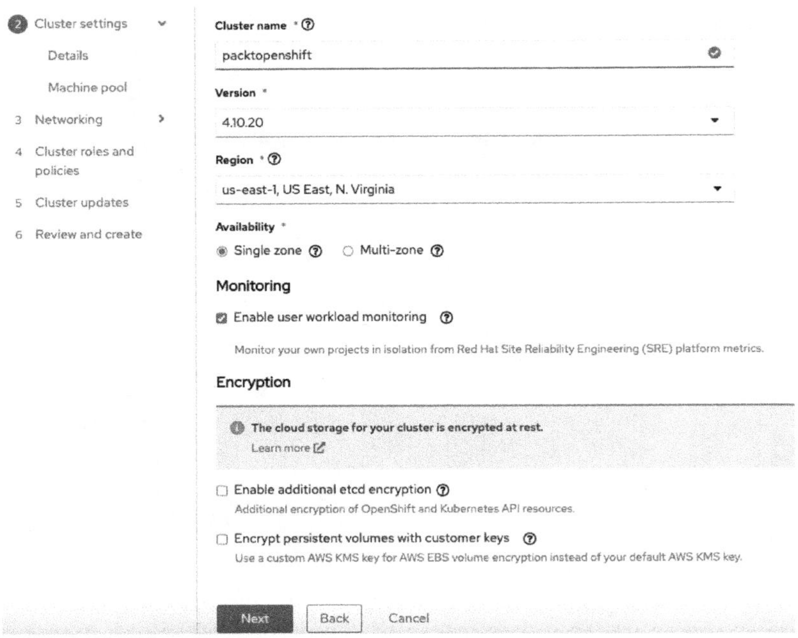

Figure 3.22 – Adding cluster details

7.   Choose your machine pool, which includes the AWS EC2 instance size and autoscaling capabilities:

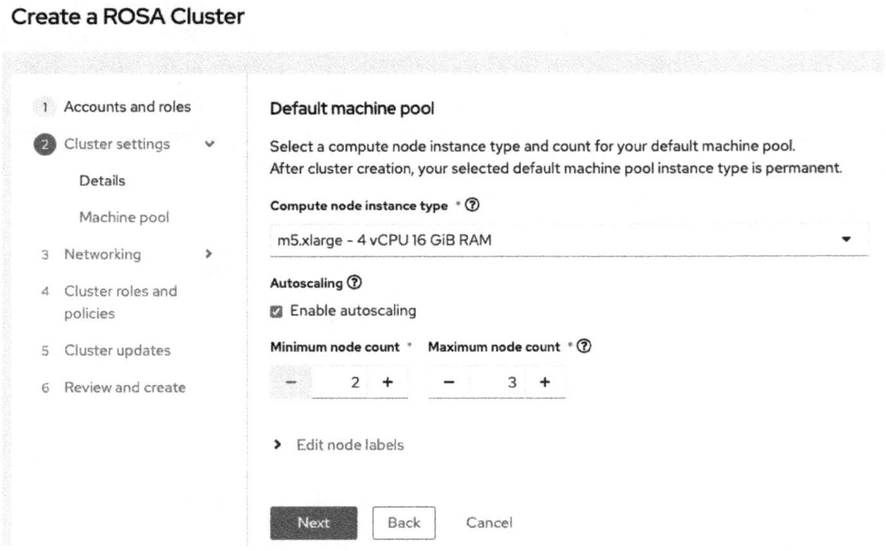

Figure 3.23 – Worker node size

8. Next are your networking options, which include whether you want the Kubernetes Control Plane/API server to be public or private, and whether you want to create a new AWS VPC for OpenShift or install the ROSA cluster into an existing VPC. Once you choose the **Virtual Private Cloud (VPC)** option, you'll have to choose the CIDR ranges for the internal Kubernetes Pod networking and cluster IP ranges:

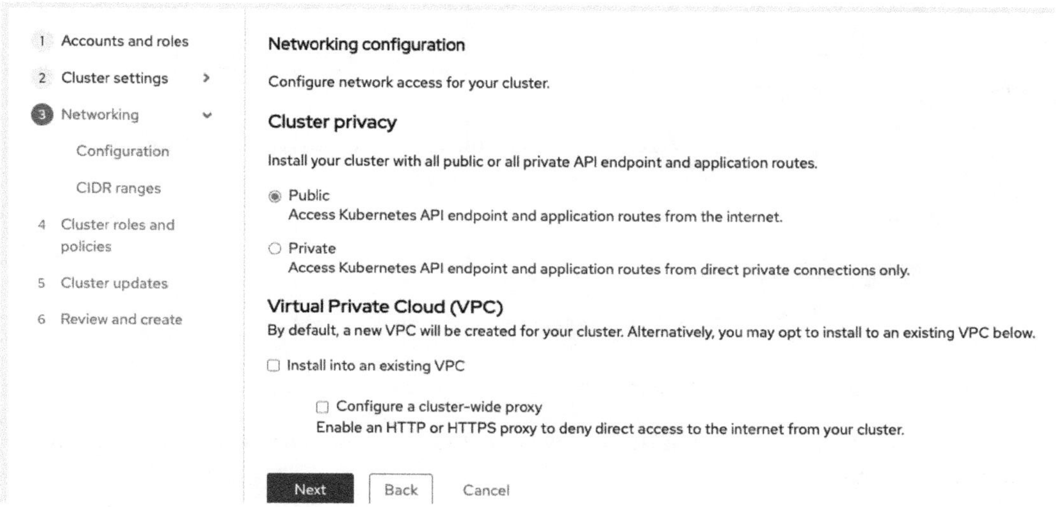

Figure 3.24 – Networking configuration

9. For cluster roles and policies, you have the option to manually set up the roles and policies or have OpenShift automatically do it for you:

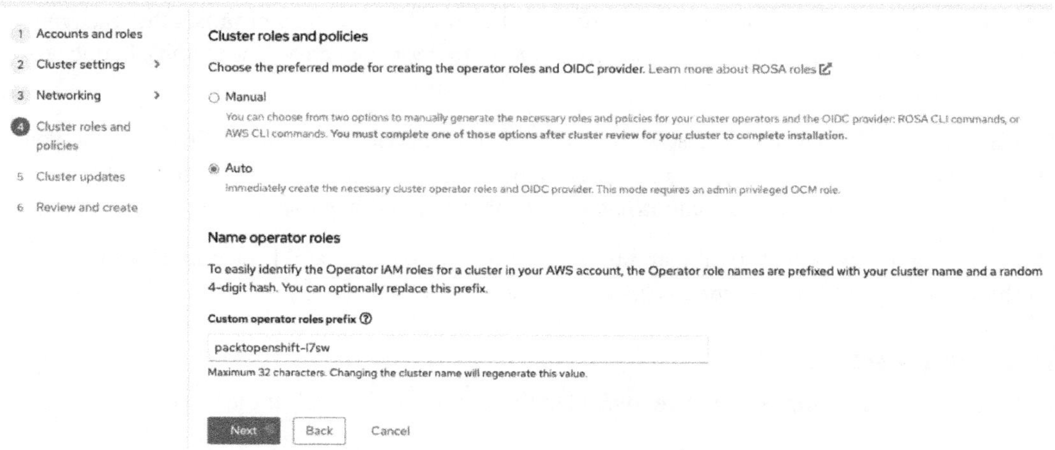

Figure 3.25 – Cluster roles and policies

10. For the last step, you can choose how you want to implement updates for the ROSA cluster. The updates that occur are based on CVE scores from the **National Vulnerability Database (NVD)**:

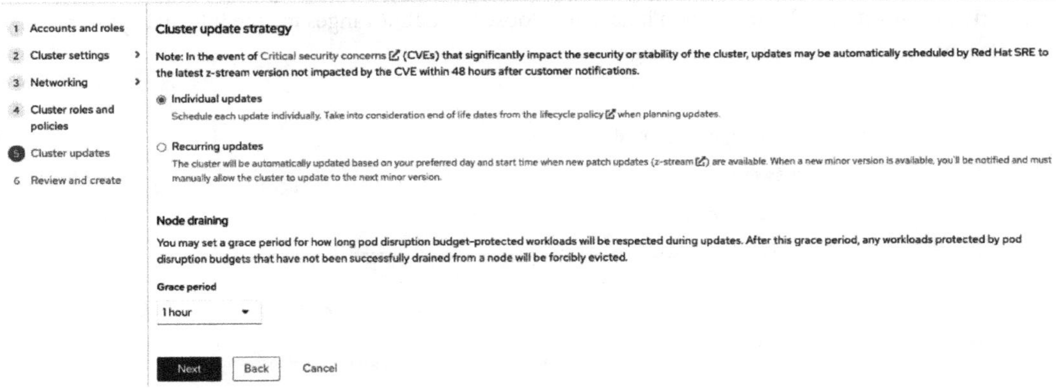

Figure 3.26 – Vulnerability scanning

11. Once you've filled in all your options, you can officially create your ROSA cluster.

Now that you know how to get up and running with OpenShift on-premises and in the cloud, let's summarize this chapter.

Summary

Regardless of what option you decide to go with when deploying Kubernetes, whether it's in a big cloud, a smaller cloud, or a PaaS solution, the goal is always the same and never changes – build an orchestration platform that can manage your containerized applications.

There are a lot of fancy tools out there, tons of different platforms, and many promises that each new and fancy platform will make your life easier from a Kubernetes perspective. The truth is, in one way or another, they all have the same goal.

The goal is to use Kubernetes to orchestrate and manage containerized applications. Ensure that as you go through each platform and tool, you have this in mind – *orchestrate my containerized applications*. If you keep that in mind, it'll make choosing and getting through the marketing fluff much easier.

In the next chapter, we'll be learning about on-premises Kubernetes and how understanding the underlying components of Kubernetes clusters is important, as well as why.

Further reading

To learn more about the topics that were covered in this chapter, take a look at the following resource:

- *Learning OpenShift*, by Denis Zuev, Artemii Kropachev, and Aleksey Usov: https://www.packtpub.com/product/learn-openshift/9781788992329

4

The On-Prem Kubernetes
Reality Check

I know what you're thinking – *On-prem? Why is this guy teaching us about on-prem Kubernetes? It's all in the cloud!*

Although it may seem like that from tech marketing and large cloud providers screaming Kubernetes at the top of their lungs, in the production world, there are a lot of on-prem Kubernetes clusters and a lot of engineers managing them. Mercedes-Benz, a popular German car manufacturer, hosts over 900 Kubernetes clusters on OpenStack, which is a private cloud solution. All those clusters are sitting in a data center, not in the public cloud. If you peel back the layers of the onion and wonder how cloud providers are running Kubernetes clusters, they're doing something similar. They have several data centers that are running Kubernetes just like you would on-prem or on a server that you can buy on eBay. The only difference is the cloud providers are running Kubernetes at a major scale around the world. However, the *how* in how Kubernetes is running isn't any different than how anyone else is running Kubernetes.

The truth is, this chapter could be an entire book – it could probably be a few books. Kubernetes, especially when it's not abstracted away in the cloud, is an extremely large topic. There's a reason why people say that Kubernetes is like a data center within itself. How and where you run on-prem Kubernetes alone is a deep topic. For example, what size infrastructure to use, how to scale your workloads, vertical and horizontal scaling, network bandwidth, high availability, and a lot more go into the conversation of what systems to use and where to run them.

By the end of this chapter, you're going to understand just how complex running on-prem Kubernetes can be, but at the same time, how rewarding it can be to an organization that's putting a lot of effort into Kubernetes. You'll have the hands-on skills and theoretical knowledge to understand how to think about scaling an organization's Kubernetes environment. One thing you'll learn from this chapter is it's a lot less about using the *cool tech* and more about thinking from an architecture perspective about how a platform team should look.

In this chapter, we're going to cover the following topics:

- Understanding operating systems and infrastructure
- Troubleshooting on-prem Kubernetes clusters
- Introducing hybrid services
- Exploring networking and system components
- Getting to know virtualized bare metal

This chapter will be a combination of hands-on and theoretical knowledge. Because we only have one chapter to cover this topic, it's safe to say that we can't show everything you'll need to know. However, this should be a good starting point for your production journey.

Technical requirements

To complete this chapter, you should first go over *Chapter 2* and *Chapter 3*. Although that might sound obvious, we want to point it out as it's crucial to understand the different deployment methods before diving into the on-prem needs of Kubernetes. Because cloud-based Kubernetes deployments abstract a lot of what you would do with on-prem, it still shows you the overall workflow of what components need to be deployed.

To work on this chapter, you should have some type of infrastructure and troubleshooting background. When it comes to on-prem Kubernetes clusters, they are extremely infrastructure-heavy, so getting through this chapter without that knowledge may be difficult. At the very least, you should have the following:

- Linux knowledge
- Server knowledge

The code for this chapter can be found in this book's GitHub repository at `https://github.com/PacktPublishing/50-Kubernetes-Concepts-Every-DevOps-Engineer-Should-Know/tree/main/Ch4`.

For the Kubeadm section of this chapter, you can follow along if you have two virtual machines available for your use. If you don't, it's perfectly fine: you can view this chapter from a theoretical perspective if that's the case. However, if you have two extra VMs available, whether they're on-prem or in the cloud, it would help you understand the overall explanations of this chapter a bit more..

Understanding operating systems and infrastructure

Everything starts at a server. It doesn't matter if you're running workloads on the cloud, on serverless platforms, or containers – everything starts at a server. The reason why engineers don't always think about servers, or where workloads start in today's world, is that the underlying infrastructure is abstracted away from us. In the cloud world, there aren't a lot of times when you'll have to ask, *what hardware are you using to run these VMs? Dell? HP?* Instead, you're worried about what happens after the servers are deployed, which is sometimes called Day-Two Ops (insert more buzzwords here). What we mean by that is instead of ordering servers online, racking them, and configuring some virtualized hypervisor on them (ESXi, KVM, Hyper-V, and so on), engineers are more concerned now with automation, application deployments, platforms, and scalability.

In many start-ups and small-to-medium-sized organizations, the typical reality is cloud computing. For larger organizations, another reality is on-prem workloads that are either virtualized or purely bare metal. If the combination of the cloud and on-prem gets brought up in discussion, this is where things such as hybrid solutions come into play, which you'll learn about later in this chapter.

Let's say you're reading this right now and you're working 100% in the cloud. You still need to understand VM sizes, scaling, the location of the VMs (the data center – regions, availability zones, geographies, and so on), network latency, and a large number of other pieces that fall into the systems and infrastructure category. For example, in the previous chapter, you learned about choosing worker node sizes for high CPU, high memory, and medium CPU/memory workloads within DigitalOcean and Linode.

In this section, you're going to learn about the core system and infrastructure needs that you must think about when architecting an on-prem Kubernetes platform.

Kubeadm Deployment

Before jumping into the theory, I wanted to showcase how you can bootstrap a Kubernetes cluster with Kubeadm. The primary reason is to show you what the process of actually deploying Kubernetes looks like while the pieces aren't abstracted away from you. Abstraction is a great thing, but it's only a great thing once you know the manual method of deployment. Otherwise, abstraction just ends up causing confusion.

For the Virtual Machines, the installation is based on Ubuntu. However, if you're using another Linux distribution, it will work, but you'll need to change the commands a bit to reflect the specific distro. For example, Ubuntu uses the Aptitude package manager and CentOS uses the Yum package manager.

Installing Control Plane and Worker Node

Let's get started.

1. First, ensure that you update Ubuntu:

    ```
    sudo apt update -y
    ```

2. Install transport layer:

```
sudo apt-get install -y apt-transport-https curl
```

3. Install Kubernetes package on Ubuntu:

```
curl -s https://packages.cloud.google.com/apt/doc/
apt-key.gpg | sudo apt-key add -
echo "deb https://apt.kubernetes.io/ kubernetes-xenial
main" | sudo tee /etc/apt/sources.list.d/kubernetes.list
```

4. Update Ubuntu again now that the Kubernetes package exists:

```
sudo apt update -y
```

5. Next, change to the root user:

```
sudo su -
```

6. Install and configure the CRI-O container runtime:

```
OS=xUbuntu_20.04
VERSION=1.22
echo "deb https://download.opensuse.org/repositories/
devel:/kubic:/libcontainers:/stable/$OS/ /" > /etc/apt/
sources.list.d/devel:kubic:libcontainers:stable.list
echo "deb http://download.opensuse.org/repositories/
devel:/kubic:/libcontainers:/stable:/cri-o:/$VERSION/$OS/
/" > /etc/apt/sources.list.d/devel:kubic:libcontainers:st
able:cri-o:$VERSION.list
curl -L https://download.opensuse.org/repositories/devel:
kubic:libcontainers:stable:cri-o:$VERSION/$OS/Release.key
| apt-key add -
curl -L https://download.opensuse.org/repositories/
devel:/kubic:/libcontainers:/stable/$OS/Release.key |
apt-key add -
```

7. Exit out of root:

```
exit
```

8. Update Ubuntu again now that CRI-O is available:

```
sudo apt update -y
```

9. Install CRI-O:

```
sudo apt install cri-o cri-o-runc -y
```

10. Reload the Daemon and enable CRI-O:

```
sudo systemctl daemon-reload
sudo systemctl enable crio --now
```

11. Check to see CRI-O is installed properly:

```
apt-cache policy cri-o
```

12. Turn off swap:

```
swapoff -a
```

13. Configure sysctl settings and ip tables:

```
sudo modprobe overlay
sudo modprobe br_netfilter


sudo tee /etc/sysctl.d/kubernetes.conf<<EOF
net.bridge.bridge-nf-call-ip6tables = 1
net.bridge.bridge-nf-call-iptables = 1
net.ipv4.ip_forward = 1
EOF
sudo sysctl --system
```

14. Install kubeadm:

```
sudo apt-get install -y kubelet kubeadm kubectl
```

The next step is configuration.

Configuring the Control Plane

We need to define variables for the kubeadm init command. This will consist of IP addresses and the Pod CIDR range. Depending on where you are deploying it, you could either have just a public subnet, or a public and private subnet.

If you have just a public subnet, use the same value for the `ip_address` and `publicIP`, along with the `CIDR` range. If you have a private and public subnet, use the public IP for the `publicIP`, the private IP for the `ip_address`, and the private IP range for the `CIDR`.

```
ip_address=10.116.0.9
cidr=172.17.0.0/16
publicIP=146.190.219.123
```

Next, initialize `kubeadm` on the Control Plane:

```
sudo kubeadm init --control-plane-endpoint $publicIP
--apiserver-advertise-address $ip_address --pod-network-
cidr=$cidr --upload-certs
```

If you are deploying in the cloud, you may find yourself in a situation where the `init` fails because the Kubelet connect communicate with the API server. This typically happens in public clouds due to network restrictions. If it happens to you, open up the following ports: `https://kubernetes.io/docs/reference/ports-and-protocols/`.

After the Kubeadm `init` is successful, you'll see a few command outputs that show how to join more Control Planes and how to join Worker Nodes. Copy the Worker Node join command and run it on the Ubuntu server that you configured as the Worker node.

Next, install the CNI.

If you don't want to use Weave, you can see the network frameworks listed here: `https://kubernetes.io/docs/concepts/cluster-administration/addons/`.

```
kubectl apply -f https://github.com/weaveworks/weave/releases/
download/v2.8.1/weave-daemonset-k8s.yaml
```

Next, we will look at system size.

System size

Considerations about the system type, size, and how many nodes will be incredibly crucial for how you decide to think about on-prem deployments. What it all comes down to is what you're planning on running on a Kubernetes cluster. If you're just starting with your first Kubernetes cluster and you want to try containerizing an application to see how it works, how the dependency works, and ultimately starting on your journey, it's going to be different than if you're running 50+ Kubernetes clusters that are running stock trading/quants applications. At the end of the day, the system size that you use will be solely based on what workload you're running.

Before you even think about creating a Kubernetes cluster on-prem, you must think about two important aspects:

- Do I have the hardware available and if not, what hardware do I have to buy?
- What type of applications am I planning on running for the next 3 to 6 months?

For example, let's say that you buy 10 servers that you're planning on running your application on. What size do the servers need to be? How will scaling work? Do you have a combination of memory-intensive apps and standard everyday apps?

Another big consideration here is *scaling*. If you're scaling horizontally, that means more Pods will be created, so more virtualized hardware will be consumed. If you're scaling vertically, that means your Pods' memory and CPU are increasing without you creating more Pods (which is known as vertical autoscaling). Not only do you have to plan for what applications you're going to be running right off the bat, but you also have to plan for how those applications will be used. If you have 500 users today and you're planning on having 2,000 users in 3 months based on company projections, that means the Pods will have an increased velocity in usage and that you may need more Pods. More usage means autoscaling, and autoscaling means more resources are needed.

Sizing considerations

The following is a list of standard sizing considerations for when you're building out a Kubernetes cluster:

- **Standard workers**: These are your everyday web server Pods or middleware that still require virtualized hardware resources, but not at the same level as more intense applications. They are your more *generic apps* if you will. The worker nodes running here are mid-to-large in terms of size. If you're just starting to get a Kubernetes cluster up and running and you're maybe moving one or two containerized apps to Kubernetes as you get going, standard workers will be just fine to get the ball rolling.

- **Memory-intensive workers**: Applications that you know will require more memory/RAM than others should be accounted for with worker nodes that contain more RAM than the standard servers that are running as worker nodes. You want to ensure that if a Pod has to scale in replicas, or more Pods are added, you have enough memory. Otherwise, Pods won't start and will stay pending until memory is allocated to them, and the scheduler can re-try scheduling the Pod for a node.

- **CPU-intensive workers**: Some applications will require more CPU and available threads to run the app. The same rules as those for memory-intensive apps apply here – if the scheduler can't schedule the Pods because there aren't enough resources available, the scheduler will wait until resources free up.

- **Special-case workers**: A special-case Pod would usually be something such as an application that's running a graphically intensive workload that needs a specific type of **Graphics Processing**

Unit (**GPU**), which means the worker node needs a dedicated GPU or an app that requires faster bandwidth, so it requires a certain type of **Network Interface Card** (**NIC**).

System location

When you run Kubernetes, there are primarily two types of nodes:

- **Control Plane:** The Control Plane is where the API server lives. Without the API server, you can't do much inside Kubernetes. It also contains the scheduler (how Pods know what node to go on), controller manager (controllers for Pods, Deployments, and so on to have the desired state), and the cluster store/database (etcd).

- **Worker node:** Worker nodes are where the Pods are installed after the Control Plane and on the worker node(s) run. They run the kubelet (the agent), container runtime (how containers run), kube-proxy (Kubernetes networking), and any other Pod that's running.

Keeping these two node types in mind, you'll have to figure out what automation techniques you want to use to run them and ultimately where/how you want to run them. It's a consideration that you shouldn't take lightly. The last thing that you want is to deploy Pods, then realize that the nodes they're running on can't handle the type of containerized app that's running. As mentioned previously, the Control Plane is where the Kubernetes API sits. The Kubernetes API is how you do everything in Kubernetes. Without it, Kubernetes wouldn't exist. With that being said, choosing where and how to run the Control Plane is the make or break between properly using Kubernetes and spending your days constantly troubleshooting.

> **Important note**
>
> Explaining the Control Plane and worker nodes could take a chapter in itself. Because this book already expects you to know Kubernetes, we're not diving into the types of Kubernetes nodes all that much. However, if you don't know about the control plane and worker nodes, we highly recommend you take the time to learn about them before continuing. A great place to start is the Kubernetes docs.

Data centers and regional locations

Data centers go down. Regions go down. **Internet Service Providers** (**ISPs**) go down. When you're architecting where you want Kubernetes to run, there are several things that you must take into consideration.

The first is where you're running. For example, if your customers are in the UK, and you decide to run your data center in New Jersey, there's going to be a ton of bandwidth issues and latency. Instead, it would make more sense to have a data center and a few co-locations throughout the UK.

Speaking of co-locations, you must make sure that you don't have a single point of failure. The reality is that data centers do go down. They have internet issues, flooding, electric issues, and outages. If that occurs, the last thing that you want is to only have a single location. Instead, you should think about, at the very least, two data center locations. If one of the data centers fails, high availability needs to be put in place to *turn on* the other data center. For example, you could have a *hot/hot* or *hot/cold* high availability scenario. *Hot/hot* is recommended as all the data is being replicated by default to the second data center. If the first data center goes down, the second data center picks up where the first left off. Another consideration is where the data centers are. If the two data centers are only 5 miles away from each other and a storm comes, both could be impacted. Because of that, you want to put some distance between data centers.

Where and how to run Kubernetes

As you saw in the previous sections, the first step to figuring out your on-prem Kubernetes infrastructure is deciding what hardware and resources you need to run the applications you're planning on running. Next, it's all about figuring out what type of worker nodes you need. It'll most likely be a combination of standard worker nodes and more intensive worker nodes with extra CPU and RAM. The final step (at least for this section) is figuring out how and where to run it.

There are several other options, but the following are a few of the popular ones:

- **OpenStack**: Although a lot of engineers in today's world think OpenStack is dead, a lot of very large organizations are still using it. For example, almost every Telco provider uses OpenStack. Mercedes-Benz (at the time of writing) is hosting over 900 (yes, 900) Kubernetes clusters running in OpenStack. OpenStack gives you the feeling of being in the cloud, but it's all on-prem and you're hosting the private cloud yourself. The Open Infrastructure Foundation has put a lot of emphasis behind running Kubernetes on OpenStack with tools such as Magnum, which is the standard for running orchestration platforms (Kubernetes, Mesos, Docker Swarm, and so on), and Loki, which is the Linux/OpenStack/Kubernetes/infrastructure stack.

- **Kubeadm**: If you don't want to go the OpenStack route and if you're using something such as a hypervisor, kubeadm is arguably the best option. There are a few different automated ways to get a Kubernetes cluster up and running, but kubeadm is more or less the most sophisticated. Using kubeadm, you can create a Kubernetes cluster that conforms to best practices. Other than installing the prerequisites on the Linux server for Kubernetes to run, kubeadm is pretty much automated. kubeadm has a set of commands that you can run that goes through several checks on the Linux server to confirm that it has all of the prerequisites and then installs the Control Plane. After that, there's an output on the terminal that gives you a command to run more Control Planes and/or run worker nodes. You copy the command from the output, paste it into another server that you're SSH'd into via the terminal, and run it. kubeadm is cool as well because it introduces you to the fact that running Kubernetes on-prem is straightforward. You can even run it on your laptop or a Rasberry Pi. There isn't a high threshold to meet to run it, especially in a Dev environment.

- **Rancher**: Rancher acts as both a Kubernetes cluster creation tool and a Kubernetes cluster manager. Within Rancher, you can create a Kubernetes cluster and host it on Rancher, create a Kubernetes cluster in the cloud, or create a raw Kubernetes cluster by provisioning Linux virtual machines. You can also manage your Kubernetes clusters from Rancher. For example, if you have a bare-metal Kubernetes cluster that's running with kubeadm or in OpenShift, you can manage it via Rancher. You can also manage cloud Kubernetes clusters.

- **Kubespray**: Kubespray, although (in our opinion) isn't the best production-level option to go for, is still an option. Kubespray uses either Ansible or Vagrant to deploy a production-ready Kubernetes cluster on virtual machines or in the cloud. Because all you need is kubeadm instead of other *middleware*, such as Vagrant or Ansible, going straight for kubeadm saves you those extra hops needed to get a cluster created. Funnily enough, Kubespray uses kubeadm underneath the hood for cluster creation (`https://github.com/kubernetes-sigs/kubespray/blob/master/docs/comparisons.md`), so that solidifies even more that there's something to say about not going the extra hops to use Kubespray and instead, just go straight to kubeadm.

Operating system

To run the Kubernetes platform, you need an operating system to run it on. The two options that you have are as follows:

- Run a bare-metal server and have the operating system run directly on the server
- Have a virtualized hypervisor, such as ESXi, that virtualizes the hardware and allows you to install the operating system on top of it

In today's world, chances are you're going to use a hypervisor. Unless there's a specific need to run bare-metal servers and run the operating system directly on the server, engineers typically opt for a hypervisor. It's much easier to manage, very scalable, and allows you to get a lot more out of the hardware.

When it comes to the operating system options, more or less, there are typically two available options. One is certainly used more than the other, but the other is gaining increased popularity.

Linux

More likely than not, you'll be running worker nodes as Linux distributions. The most popular battle-tested distributions are Red Hat, CentOS, and Ubuntu. Linux is usually the *out-of-the-box* solution when it comes to Kubernetes worker nodes, and at the time of writing this book, you can only run Kubernetes Control Planes on Linux servers.

Windows

Although not seen all that much, especially with open-sourced and cross-platform versions of .NET, you can run Windows Server as a Kubernetes worker node. If you want to run Windows Server as a worker node, there are a few considerations. First, you must be running Windows Server LTSC 2019 or above. At the time of writing this book, the two options available are Windows Server 2019 and Windows Server 2022.

With the Windows Server option, you will have to buy licenses and keep **Client Access License (CAL)** considerations in mind.

Now that you have an understanding of the overall operating system and infrastructure components of an on-prem Kubernetes cluster, in the next section, you'll learn how to troubleshoot the environment you're building.

Troubleshooting on-prem Kubernetes clusters

If you come from a systems administration/infrastructure background, troubleshooting Kubernetes clusters is going to come to you pretty naturally. At the end of the day, a Kubernetes cluster consists of servers, networking, infrastructure, and APIs, which are essentially what infrastructure engineers are working on day to day.

If you're a developer, some of these concepts may be new to you, such as troubleshooting networks. However, you'll be very familiar with a few troubleshooting techniques as well, such as looking at and analyzing logs.

The whole idea of troubleshooting a Kubernetes cluster is to look at two pieces:

- The cluster itself
- The Pods running inside the cluster

The cluster itself, including networking, servers, operating systems, and scalability, is going to be thought of from more of an infrastructure perspective, where something such as the **Certified Kubernetes Administrator (CKA)** comes into play nicely. The Pods, Deployments, container errors, and Pods not starting properly are going to be thought of more from a developer perspective, so learning the concepts of the **Certified Kubernetes Application Developer (CKAD)** would be a great stepping stone.

In this section, you're going to learn about the key ways to think about troubleshooting Kubernetes clusters and how to figure out problems in a digestible way.

Server logs and infrastructure troubleshooting

Although there's an entire chapter in this book that goes over logging and observability (*Chapter 7*) let's talk about logging in a cluster sense. Typically, when you're working with any type of observability metrics, such as logging, in the Kubernetes world, engineers are primarily thinking about the logs for an application. Those logs will help them troubleshoot a failing app and figure out what happened in the first place. However, from a Kubernetes on-prem perspective, server logging is a crucial piece.

For the most part, unless otherwise specified, all the logs from the Control Plane and worker nodes typically go into `/var/log` on the Linux server. For Control Planes, the logs are at the following paths:

- `/var/log/kube-apiserver.log`
- `/var/log/kube-scheduler.log`
- `/var/log/kube-controller-manager.log`

For worker nodes, the logs are at the following paths:

- `/var/log/kubelet.log`
- `/var/log/kube-proxy.log`

Out of the box, there isn't a specific logging mechanism that Kubernetes uses. That's essentially up to you to decide. The two primary ways that engineers capture logs for clusters are as follows:

- Use a node logging agent that runs on every node across the cluster
- Have a log aggregator capture the logs from `/var/log` and send them to a logging platform

You can find more documentation on troubleshooting clusters at `https://kubernetes.io/docs/tasks/debug/debug-cluster/`.

Network observability

The networking piece of a Kubernetes cluster, which you'll learn about shortly, is an extremely complex piece of Kubernetes within itself. Networking inside of Kubernetes is just as important as the Kubernetes API and all the other pieces that make up Kubernetes.

The two things that you want to look out for are **cluster latency** and **Pod latency**. With cluster latency, it's most likely going to come down to the standard systems administration troubleshooting around checking bandwidth, QoS on routers, how many packets are getting pushed through, NICs, and more. From a Pod latency perspective, it'll most likely start at the cluster level in terms of the issues that the cluster may be having, but to troubleshoot, you'll most likely look into something such as a service mesh.

Service mesh is a huge topic in itself, which could probably cover an entire course/book, but you'll learn how to get started with it in the *Exploring networking and system components* section.

Kubernetes metrics

Most resources created in Kubernetes (Deployments, Pods, Services, and so on) have a metrics endpoint that can be found via the `/metrics` path when making an API call on Kubernetes resources.

The metrics server collects logs from kubelets, which are agents that run on each Kubernetes node, and exposes them to the Kubernetes API server through the metrics API. However, this is primarily used for autoscaling needs. For example, the metrics will tell Kubernetes, *Hey, Pods are running low on memory utilization; we need a new Pod to handle application utilization.* Then, the vertical or horizontal autoscaler will kick off and do its job to create a new Pod, or vertically scale the current Pod(s).

If you want to collect metrics for, say, the monitoring platform that you use for observability, you'd want to collect metrics from the /metrics/resource/resource_name kubelet directly. Many observability platforms such as Prometheus will ingest these metrics and use them for troubleshooting and performance troubleshooting.

crictl

Inside of every Kubernetes cluster, specifically running on the worker nodes, is a container runtime. Container runtimes such as containerd and CRI-O are mostly used in Kubernetes environments. Those container runtimes help make containers and Pods run. Because of that, it's important to ensure that the container runtime is working as expected.

crictl helps you troubleshoot the container runtime. You can run a few commands directed at a Pod that'll help you understand what's happening inside of a container. Keep in mind that crictl is in beta, but it's still a great troubleshooting tool.

In the following example, crictl is listing a set of Pods:

```
crictl pods
```

Then, it can look inside each Pod to see the containers that are running:

```
crictl pods -name pod_name
```

You can also list out containers that are running and bypass Pods:

```
crictl ps -a
```

You can find more information about crictl at https://github.com/kubernetes-sigs/cri-tools/blob/master/docs/crictl.md.

kubectl

Wrapping up this section, let's talk about the kubectl command, which is the typical way that engineers interact with Kubernetes via the CLI. You'll see three primary commands for troubleshooting:

* kubectl describe: This command tells you the exact makeup of the Kubernetes Deployment, such as how it's running, what containers are running inside of it, ports that are being used, and more. This is a great first step to understanding what could be going wrong inside of a Deployment.

For example, if you had a Deployment called `nginx-deployment`, you'd run the following command:

```
kubectl describe deployment nginx-deployment
```

The following output showcases how `describe` looks for Deployments.

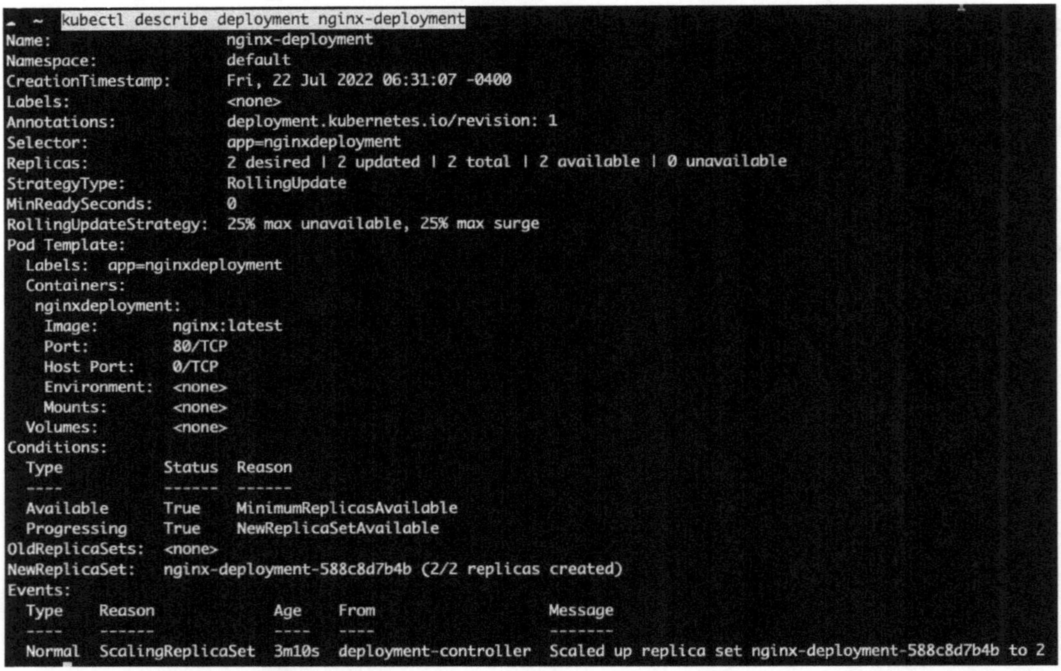

```
~ kubectl describe deployment nginx-deployment
Name:                   nginx-deployment
Namespace:              default
CreationTimestamp:      Fri, 22 Jul 2022 06:31:07 -0400
Labels:                 <none>
Annotations:            deployment.kubernetes.io/revision: 1
Selector:               app=nginxdeployment
Replicas:               2 desired | 2 updated | 2 total | 2 available | 0 unavailable
StrategyType:           RollingUpdate
MinReadySeconds:        0
RollingUpdateStrategy:  25% max unavailable, 25% max surge
Pod Template:
  Labels:  app=nginxdeployment
  Containers:
   nginxdeployment:
    Image:        nginx:latest
    Port:         80/TCP
    Host Port:    0/TCP
    Environment:  <none>
    Mounts:       <none>
  Volumes:        <none>
Conditions:
  Type           Status  Reason
  ----           ------  ------
  Available      True    MinimumReplicasAvailable
  Progressing    True    NewReplicaSetAvailable
OldReplicaSets:  <none>
NewReplicaSet:   nginx-deployment-588c8d7b4b (2/2 replicas created)
Events:
  Type    Reason             Age    From                    Message
  ----    ------             ----   ----                    -------
  Normal  ScalingReplicaSet  3m10s  deployment-controller   Scaled up replica set nginx-deployment-588c8d7b4b to 2
```

Figure 4.1 – Kubernetes Deployment output

- `kubectl cluster-info dump`: This command is a literal dump of every single thing that's happened on the cluster that was recorded. By default, all of the output is sent STDOUT, so you should ideally send the output to a file and look through it as it's extremely verbose with a lot of data.

The following screenshot has been cut off for simplicity, but it's an example of the information shown with the `kubectl cluster-info dump` command:

```
I0608 12:42:34.194982        1 controller.go:884] Started provisioner controller k8s.io/minikube-hostpath_minikube_f128d371-5449-
!
==== END logs for container storage-provisioner of pod kube-system/storage-provisioner ====
{
    "kind": "EventList",
    "apiVersion": "v1",
    "metadata": {
        "resourceVersion": "1851361"
    },
    "items": [
        {
            "metadata": {
                "name": "nginx-deployment-588c8d7b4b-wmg9z.1704201c144e865e",
                "namespace": "default",
                "uid": "6130e548-9297-410a-b69b-692254752795",
                "resourceVersion": "1851039",
                "creationTimestamp": "2022-07-22T10:31:07Z"
            },
            "involvedObject": {
                "kind": "Pod",
                "namespace": "default",
                "name": "nginx-deployment-588c8d7b4b-wmg9z",
                "uid": "06fb02a9-64ca-47ac-aeca-bfb3e9bbb552",
                "apiVersion": "v1",
                "resourceVersion": "1851033"
            },
            "reason": "Scheduled",
            "message": "Successfully assigned default/nginx-deployment-588c8d7b4b-wmg9z to minikube",
            "source": {
                "component": "default-scheduler"
            },
            "firstTimestamp": "2022-07-22T10:31:07Z",
            "lastTimestamp": "2022-07-22T10:31:07Z",
            "count": 1,
            "type": "Normal",
            "eventTime": null,
            "reportingComponent": "",
            "reportingInstance": ""
```

Figure 4.2 – Cluster dump output

- `kubectl logs`: This command is the bread and butter to understanding what's happening inside of a Pod. For example, let's say that you have a Pod called `nginx-deployment-588c8d7b4b-wmg9z`. You can run the following command to see the log output for the Pod:

```
kubectl logs nginx-deployment-588c8d7b4b-wmg9z
```

The following screenshot shows a sample of what logs look like for Pods.

```
 ~ kubectl logs nginx-deployment-588c8d7b4b-wmg9z
/docker-entrypoint.sh: /docker-entrypoint.d/ is not empty, will attempt to perform configuration
/docker-entrypoint.sh: Looking for shell scripts in /docker-entrypoint.d/
/docker-entrypoint.sh: Launching /docker-entrypoint.d/10-listen-on-ipv6-by-default.sh
10-listen-on-ipv6-by-default.sh: info: Getting the checksum of /etc/nginx/conf.d/default.conf
10-listen-on-ipv6-by-default.sh: info: Enabled listen on IPv6 in /etc/nginx/conf.d/default.conf
/docker-entrypoint.sh: Launching /docker-entrypoint.d/20-envsubst-on-templates.sh
/docker-entrypoint.sh: Launching /docker-entrypoint.d/30-tune-worker-processes.sh
/docker-entrypoint.sh: Configuration complete; ready for start up
2022/07/22 10:31:12 [notice] 1#1: using the "epoll" event method
2022/07/22 10:31:12 [notice] 1#1: nginx/1.23.1
2022/07/22 10:31:12 [notice] 1#1: built by gcc 10.2.1 20210110 (Debian 10.2.1-6)
2022/07/22 10:31:12 [notice] 1#1: OS: Linux 5.10.104-linuxkit
2022/07/22 10:31:12 [notice] 1#1: getrlimit(RLIMIT_NOFILE): 1048576:1048576
2022/07/22 10:31:12 [notice] 1#1: start worker processes
2022/07/22 10:31:12 [notice] 1#1: start worker process 32
2022/07/22 10:31:12 [notice] 1#1: start worker process 33
2022/07/22 10:31:12 [notice] 1#1: start worker process 34
2022/07/22 10:31:12 [notice] 1#1: start worker process 35
```

Figure 4.3 – Nginx Pod log output

Regardless of where a Kubernetes cluster is running, you're always going to have to troubleshoot certain aspects of it. The tips in this section should help in an on-prem, and even sometimes a cloud, scenario.

Introducing hybrid services

From 2014 to 2015, what most organizations and engineers alike were reading sounded something similar to *data centers will go away*, *the cloud is the future*, and *everyone that isn't in the cloud will be left behind*. Organizations started to feel pressured to move to the cloud and engineers started to get nervous because the skills they had honed for years were becoming obsolete. Coming back to the present, which is 2022 at the time of writing this book, mainframes still exist... so, yes, many organizations are still running on-prem workloads. Engineers that have an infrastructure and systems background are doing quite well for themselves in the new cloud-native era. The reason why is that 100% of the skills they have learned, other than racking and stacking servers, are still very relevant for the cloud and Kubernetes.

In the *Understanding operating systems and infrastructure* section, you may remember reading about on-prem workloads and how they're still relevant in today's world. Although tech marketing may be making you feel otherwise, the truth is that on-prem workloads are still very much used today. They're used so much that organizations such as AWS, Microsoft, and Google are realizing it, and they're building services and platforms to support the need for a true hybrid environment, which means using on-prem and cloud workloads together, often managed in the same location.

In this section, you're going to learn about the major cloud provider hybrid services, along with a few other companies that are helping in this area.

Azure Stack HCI

Azure Stack HCI is the hybrid cloud offering from Microsoft. It gives you the ability to connect your on-prem environment to Azure. Azure Stack HCI typically comes running inside of a server from a vendor, though you can install it yourself on a compatible server with compatible hardware. It installs similar to any other operating system, but there's a lot of complexity around the requirements. A few of the requirements include the following:

- At least one server with a maximum of 16 servers
- Required to deploy to two different sites
- All servers must have the same manufacturer and use the same model
- At least 32 GB of RAM
- Virtualization support on the hardware (you have to turn this on in the BIOS)

You can dive into the requirements a bit more; you'll find that it goes pretty in-depth. From a time perspective, you're probably better off buying an Azure Stack HCI-ready server from a vendor.

An interesting part of Azure Stack HCI is underneath the hood – it's pretty much just Windows Server 2022 running Windows Admin Center. Because of that, you could completely bypass Azure Stack HCI and do the following instead:

1. Run a bunch of Windows Server 2022 Datacenter servers.
2. Cluster them up.
3. Install Windows Admin Center.
4. Connect the servers to Azure.
5. Run AKS on the servers.

Google Anthos

Anthos is arguably the most mature hybrid cloud solution that's available right now. There are a ton of ways to automate the installation of Anthos with, for example, Ansible, and the hardware requirements to get it up and running are far more lite (at the time of writing this book) compared to Azure Stack HCI.

The hardware requirements are as follows:

- Four cores for CPU minimum, Eight cores recommended
- 16 GB of RAM minimum, 32 GB recommended
- 128 GB storage minimum, 256 GB recommended

Like Azure Stack HCI, Anthos runs on-prem in your data center and connects to GCP to be managed inside of the GCP UI or with commands/APIs for GCP. The goal here is to run GCP to manage Kubernetes clusters on-prem and in the cloud.

A quick note about other infrastructure managers

Although perhaps not considered hybrid cloud in itself as a platform, there are a few platforms that help you manage workloads anywhere. Two of the primary ones at the time of writing are as follows:

- **Azure Arc**: Azure Arc, as the name suggests, requires an Azure subscription. However, the cool thing about it is that you can manage Kubernetes clusters anywhere. If you have Kubernetes clusters in, for example, AWS, you can manage them with Azure Arc. If you have Kubernetes clusters on-prem, you can manage them with Azure Arc.

- **Rancher**: Rancher is a vendor-agnostic solution that does all the management goodness that Azure Arc does, with a few other key features such as logging, deployments of Kubernetes servers, and security features to help you fully manage your Kubernetes clusters that are running anywhere.

In the next section, you'll learn about the overall network administration that's needed inside of a Kubernetes cluster.

Exploring networking and system components

Networking in a Kubernetes cluster, aside from the Kubernetes API itself, is what makes Kubernetes truly *tick*. Networking comes into play in various ways, including the following:

- Pod-to-Pod communication

- Service-to-Service communication

- How nodes talk to each other inside of the cluster

- How users interact with your containerized applications

Without networking, Kubernetes wouldn't be able to perform any actions. Even from a control plane/ worker node perspective, worker nodes can't successfully communicate with control planes unless proper networking is set up.

This section could be, at the very least, two chapters in itself. Because we only have one section to hammer this knowledge down, let's talk about the key components.

kube-proxy

When you first start to learn about how networking works inside of Kubernetes and how all resources communicate with each other, it all starts with kube-proxy. kube-proxy is almost like your switch/router in a data center. It runs on every node and is responsible for local cluster networking.

It ensures the following:

- That each node gets a unique IP address
- It implements local iptables or IPVS rules
- It handles the routing and load balancing of traffic
- It enables communication for Pods

In short, it's how every resource in a Kubernetes cluster communicates.

CNI

The first step is kube-proxy, but to get it deployed, it needs to have a backend. That *backend* is the **Container Network Interface** (**CNI**). Attempting to run kube-proxy without a CNI is like trying to run a network on Cisco without having Cisco equipment – it doesn't work.

The CNI is a network plugin, sometimes called a network framework, that has the responsibility of inserting a network framework into a Kubernetes cluster to enable communication, as in, to enable kube-proxy.

There are a ton of popular CNIs, including the following:

- Weave
- Flannel
- Calico

Kubernetes resource communication

When you deploy Pods, especially microservices, which are X number of Pods running to make up one application, you need to ensure that Pod-to-Pod communication and Service-to-Service communication work.

Pods communicate with each other via an IP address, which is given by kube-proxy. The way that services communicate with each other is by hostname and IP address, which is given by CoreDNS. Services provide a group of Pods associated with that service with a consistent DNS name and IP address. CoreDNS ensures the translation from hostnames to IP addresses.

DNS

Under the hood, Kubernetes runs **CoreDNS**, a popular open source DNS platform for converting IP addresses into names. When a Pod or a Service has a DNS name, it's because the CoreDNS service (which is a running Pod) is running on Kubernetes properly.

Service mesh and Ingress

Much like a lot of the other topics in this chapter, service meshes could be an entire chapter – the two topics mentioned here could be an entire book. However, let's try to break it down into one section.

An Ingress controller lets you have multiple Kubernetes Services being accessed via one controller or load balancer. For example, you could have three Kubernetes services named App1, App2, and App3, all connected to the same Ingress controller and accessible over the `/app1`, `/app2`, and `/app3` paths. This is possible via routing rules, which are created for the Ingress controller.

A service mesh, in short, helps you encrypt east-west traffic or service-to-service traffic, and troubleshoot network latency issues.

Sometimes, depending on the service mesh that you use, you may not need an Ingress controller as the service mesh may come built with one.

For Ingress controllers, check out Nginx Ingress, Traefik, and Istio. For service meshes, check out Istio.

In the next section, you're going to learn about the ins and outs of how to think about virtualized bare metal and a few vendors that help on this journey.

Getting to know virtualized bare metal

If/when you're planning to run Kubernetes on-prem, two questions may pop up:

- Where are we going to run Kubernetes?

- How are we going to run Kubernetes?

In today's world, chances are you're not going to run Kubernetes directly on bare metal (although you could, and some companies do). You'll probably run Kubernetes on a hypervisor such as ESXi or in a private cloud such as OpenStack. You may also run Kubernetes on virtualized bare metal, which is different than running it on a hypervisor.

In this section, let's learn what virtualized bare metal is and a few ways that you can run it.

Virtualizing your environment

When thinking about virtualized bare metal, a lot of engineers will most likely think about a hypervisor such as VMware's ESXi or Microsoft's Hyper-V. Both are great solutions and allow you to take a bare-metal server that used to only be able to run one operating system and run multiple operating systems. There are many other pieces to a virtualized environment, such as virtualized hardware, networking, and more, all of which are extensive topics and could be an entire book in itself.

This not only helps you use as many resources out of the server as you can, but it also allows you to save on cost because servers are expensive.

The other solution is to run as close to bare metal as possible, but you don't host it. Instead, you *rent* bare-metal servers from a hosting provider. When you rent the servers, they give you SSH or RDP access and you can access them the same way that you would if the servers were running in your data center. There's a UI that you can use to create the servers, maybe some automated ways to do so if the platform allows it, and you can create Windows and/or Linux servers like you would on any other platform, such as if you were creating a web service or a server to host applications.

Where to run Kubernetes

When thinking about where you'd want to run ESXi or Hyper-V, that'll most likely come down to what servers you currently have, the partnerships with vendors your business has, and what resources (CPU, RAM, and so on) you need on the servers.

When it comes to the "*as close to bare metal as possible*" option, although there are many vendors, two stick out in the Kubernetes space:

- **Equinix**: Equinix allows you to – not only from a UI perspective but also from an automation perspective – use tools such as Terraform to create virtualized bare-metal servers for both Linux and Windows distributions. You can also manage networking pieces such as BGP and other routing mechanisms, as well as use on-demand, reserved, and spot servers:

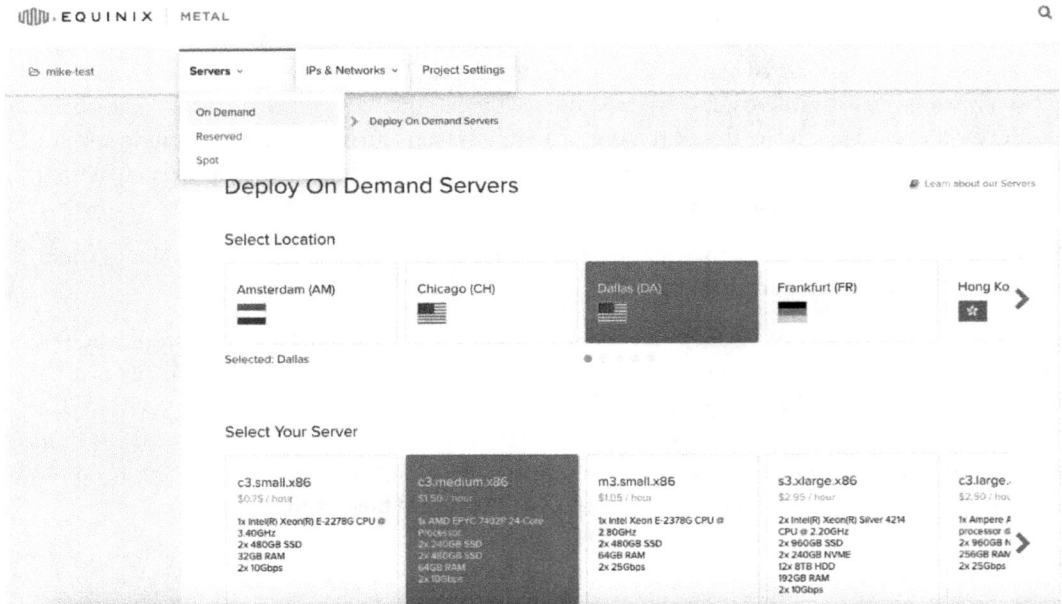

Figure 4.4 – Equinix metal server creation

In the following screenshot, you can see the general starting point in Equinix to start deploying servers:

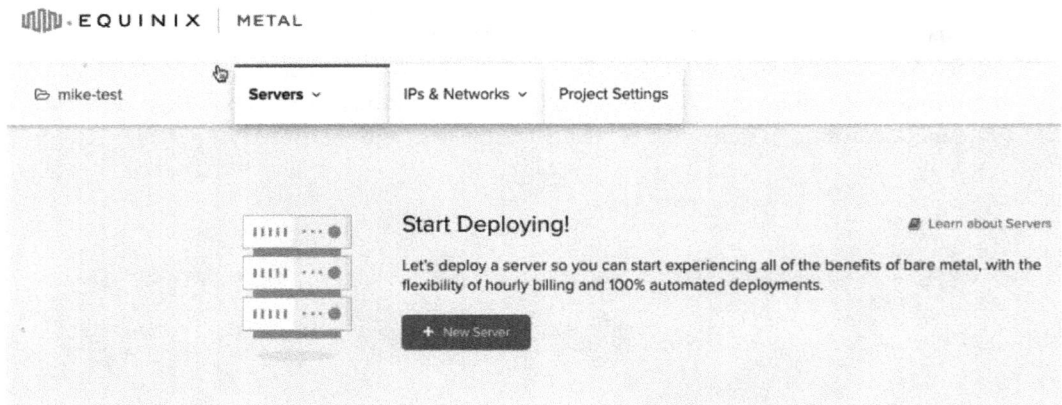

Figure 4.5 – Equinix Metal deployment page

- **OpenMetal**: OpenMetal is a full-blown virtualized bare-metal solution for running OpenStack. One of the cool parts about OpenMetal is that you get true SSH access to the literal servers that are running OpenStack, so you have a ton of flexibility and customization options, just like you would in any OpenStack environment. The only thing you don't have access to is the actual hardware itself as that's managed by OpenMetal, but you most likely don't want access to it anyway if you're looking for a virtualized bare-metal solution:

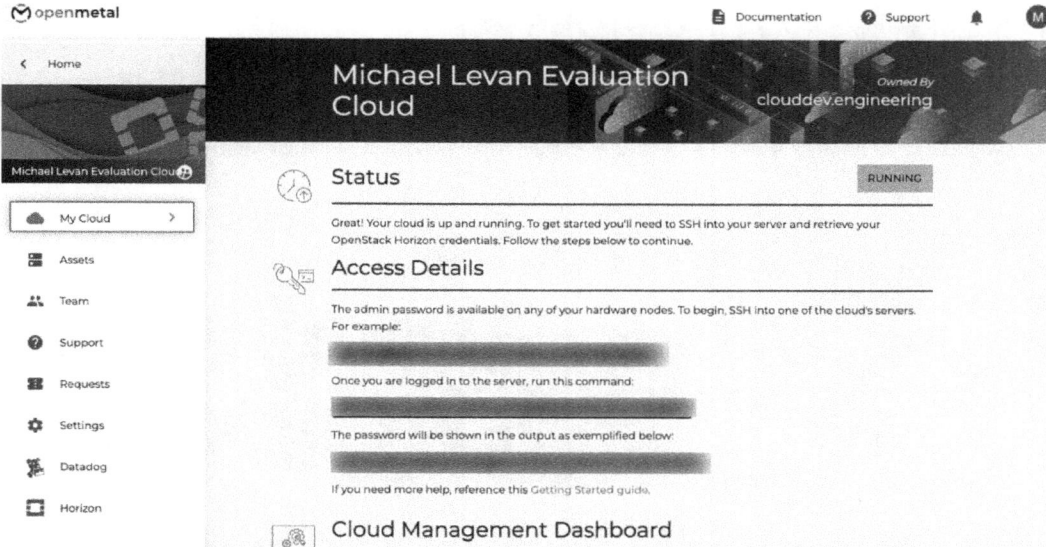

Figure 4.6 – OpenMetal dashboard

The following screenshot shows the standard UI in OpenStack, which is running on OpenMetal. This shows that nothing is different from using OpenStack on any other environment, which is great for engineers that are already used to it:

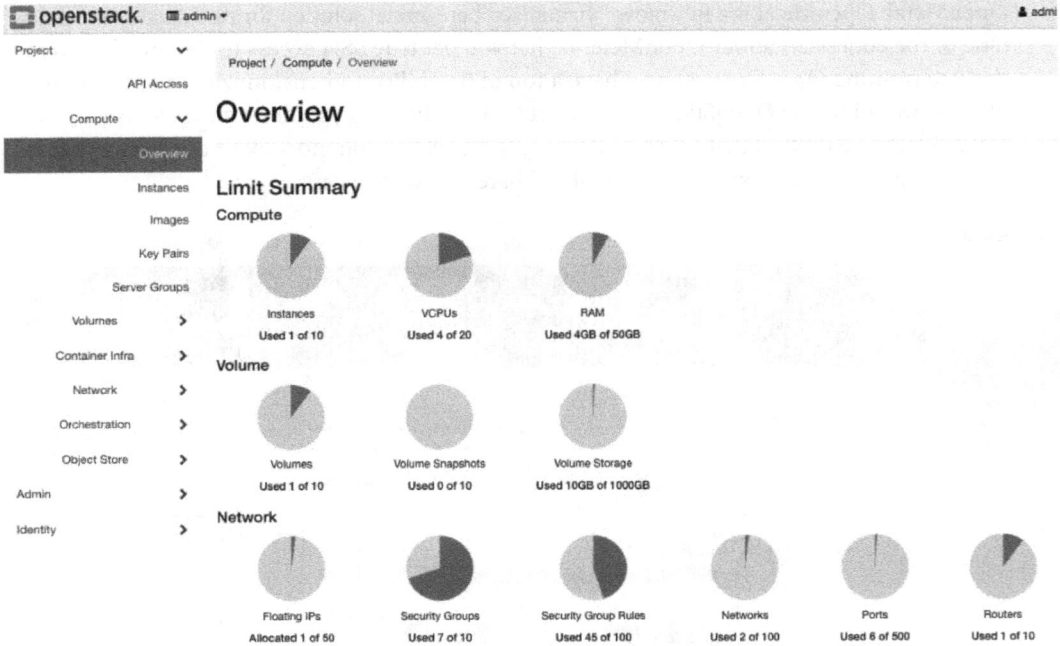

Figure 4.7 – OpenStack's Overview dashboard

If you're interested in running Kubernetes on-prem, but still want the feel of a *cloud*-based environment, OpenStack running on OpenMetal is a great place to start.

Summary

There's a lot that wasn't talked about in this chapter – storage, different interface types, hardware types, the ins and outs of Kubernetes clusters, and a lot more. The reason why is that this chapter could only be so long and a lot of the topics could take up an entire chapter.

However, the goal of this chapter was to give you a place to start.

As you learned throughout this chapter and may have come to realize, managing Kubernetes on-prem can almost feel like an entire data center within itself. You have networking concerns, scalability concerns, storage concerns, network concerns… the list goes on and on. However, if you want the flexibility of managing Kubernetes yourself without relying on a cloud provider, then this chapter went over what you should think about from the beginning.

Running Kubernetes on-prem is no easy task. You will most likely have to have a team of engineers – or at the very least two to three engineers with a very strong systems administration and network administration background. If you don't already have those skills, or if your team doesn't, this chapter should have given you a good starting point on where to look.

In the next chapter, you'll start looking at the *how* and *why* behind deploying applications on Kubernetes.

Further reading

To learn more about the topics that were covered in this chapter, take a look at the following resource:

- *OpenStack Cookbook*, by Kevin Jackson, Cody Bunch, and Egle Sigler: `https://www.packtpub.com/product/openstack-cloud-computing-cookbook-fourth-edition/9781788398763`

Part 2:
Next 15 Kubernetes Concepts – Application Strategy and Deployments

Kubernetes cluster management is such a huge part of managing Kubernetes as a whole. In fact, that's probably why the **Certified Kubernetes Administrator** (**CKA**) program is so popular and a prerequisite to the **Certified Kubernetes Security** (**CKS**) exam. Although certifications aren't the be-all-and-end-all, and you definitely don't need them to work with Kubernetes, it gives you an idea of what a critical component is, which is the infrastructure and overall cluster management.

The second huge piece, which is incredibly important and absolutely needed, is the containerized application deployment piece. The second set of the 50 concepts will go into overall deployments, which you'll learn about in these sections.

Typically, when an engineer first gets started with Kubernetes, they'll deploy a simple web app such as nginx or Hello World. They'll grab the container image, put it in a Kubernetes manifest, run `kubectl apply -f`, and call it a day. The point is that popping a container image in a manifest and running a command is the easy part. The hard part is thinking about the actual strategy that you're utilizing. Will you implement GitOps? CI/CD? A service mesh? How are you managing automation and repeatability? How many apps need to be deployed? There's a lot that goes into the thought process from a strategy and architecture standpoint.

By the end of these, you should have an understanding of the various ways to deploy containerized apps. Although we couldn't go into every specific use case, you should be able to take what's in these two chapters and have a good path forward for what to use in your production environment.

This part of the book comprises the following chapters:

- *Chapter 5, Deploying Kubernetes Apps Like a True Cloud Native*
- *Chapter 6, Kubernetes Deployment – Same Game, Next Level*

5

Deploying Kubernetes Apps Like a True Cloud Native

When engineers start hearing about Kubernetes or want to start implementing it, the typical reason is from a Dev perspective of managing and deploying applications. The whole premise around Kubernetes making engineering teams' lives easier, regardless of whether it's Dev or Ops, is based on application deployment.

Deploying applications is at the forefront of every business's mind, whether it's a website, some mobile application, or an internal app in any company, from a software company to an auto-parts company to a beer manufacturer. Regardless of the industry, almost every company deploys some type of application and some type of software. As all engineers know, deploying and maintaining an application successfully isn't an easy task. Whether you're running an app on bare metal, on a VM, in the cloud, or even in a container, that app could be (and most likely is) the make or break between a successful business and a bankrupt company.

Throughout this chapter, you're going to notice that a lot of topics covered will remind you of how other application deployments work. From the actual deployment to scaling and upgrading, the overall concepts are the same. For example, scaling an application is scaling an application. There's no magical new methodology with Kubernetes. However, what Kubernetes does give you is the ease of scalability. With that being said, the major thing you'll notice throughout this chapter is that Kubernetes isn't reinventing the wheel. It's making what we've been doing for 30+ years easier.

In this chapter, we're going to cover the following main topics:

- Cloud-native apps
- Controllers, controller deployments, and Pods
- Segregation and namespaces
- Stateless and stateful apps
- Upgrading deployments

Technical requirements

To follow along with this chapter, you should have already deployed a Kubernetes app via a Kubernetes manifest. This chapter is going to break down the process of things such as deploying apps and what a Kubernetes manifest is, but to fully grasp the chapter, you should be familiar with the deployment process. Think of it like this – you should be at a beginner/mid level with the Kubernetes deployment process, and this chapter will get you to the production level.

The code for this chapter can be found at the following GitHub URL: `https://github.com/PacktPublishing/50-Kubernetes-Concepts-Every-DevOps-Engineer-Should-Know/tree/main/Ch5`.

Understanding cloud-native apps

Although the whole *cloud-native* thing can feel a bit buzzword-ish in today's world, there is some merit behind the idea of building cloud-native apps. The way an application is architected matters as it relates to how it can be deployed, managed, and maintained later. The way a platform is built matters because that's the starting point for how an application can be deployed and how it can be run.

Throughout the years of technology's existence, there have been multiple different methodologies around how applications are architected and built. The original methods were formed around on-premises systems, such as mainframes and servers. After that, applications started to be architected for virtualized hardware platforms, such as ESXi, and other virtualization products, with the idea in mind of utilizing more of the server, but for different workloads instead of just one workload running like in the bare-metal days. After virtualization, there was architecture and planning for apps around cloud workloads, which started to introduce the idea of cloud native and how applications would work if they only ran in the cloud. Considerations such as bandwidth, size of servers, and overall cost consumed a lot of conversations around cloud workloads, and still do.

Now, we're faced with the *fourth phase*, which is containerized workloads. Containerized workloads really kicked off the focus around cloud-native applications and deployments, and for good reason, especially with the idea of microservices starting to become a real thing for many organizations that would've thought it wasn't possible just 5 years ago.

In this section, you're going to learn what cloud-native applications are and a brief history of application architecture, cloud deployments, and microservices.

What's a cloud-native app?

Before deploying applications in a cloud-native way, let's take a step back and think about a core computer science concept – distributing computing.

Distributed computing is a field that studies distributed systems, and distributed systems are systems that have components located on different network-connected computers. Those different network-connected computers then communicate with each other to send data, or packets, back and forth.

The important part here is this – distributed systems equal multiple software components that are on multiple systems but run as a single system. Distributed computing sounds like microservices, right? (More on microservices later.)

Cloud native takes the concept of distributed computing and expands it to a whole other level. Think about it from an AWS or Azure perspective. AWS and Azure are by definition distributed systems. When you log in to the AWS portal, there are a ton of services at your fingertips – EC2, databases, storage, and a lot more. All of those services that you interact with are from a *single system*, but the network components that make up the *single system* span hundreds of thousands of servers, across multiple data centers across the entire world. Cloud native doesn't just mean the public cloud, however. Remember, the cloud is a distribution of services. Something that's *cloud native* can also be, for example, an entire OpenStack server farm.

Combining the whole idea of cloud native and distributed computing, you have a major concept – cloud-native applications. Cloud-native apps aim to give you the ability to design and build apps that are the following:

- Easily scalable

- Resilient

- Elastic

The important thing to remember is that these concepts aren't any different than what we've already had in the engineering world. We've had distributed computing for a long time. We've had distributed applications for a while. What we didn't always have is the ability to easily implement distributed computing. Thinking about the AWS or Azure example from previously, how long would it take us to build the same infrastructure as Azure or AWS? Then, think about how many people it would take to manage and maintain it. With distributed computing at the cloud level, all of the *day-one* configurations are abstracted away from you, leaving you with only worrying about the *day-two* complexities of building a cloud-native/distrusted computing application.

In the cloud, if you want to scale your application, you click a few buttons, write a few lines of code, and boom, you have autoscaling groups. If you want to build resilient applications, you point and click on what data centers you want your apps to run in instead of having to physically build out those data centers. Again, the concept of distributed computing is the same thing as cloud native and cloud-native apps. The difference is that you don't have to worry about building out the data center. You just have to worry about scaling the app.

Cloud-specific cloud native

One major point to keep in mind, whether it's with a standard app deployment in the cloud or an app deployment in Kubernetes, is *cloud native* doesn't just mean *the cloud*. It's more or less the overall concept, but again, the whole idea of cloud native is distributed computing without the need to focus on the day-one implementation and configuration.

For example, let's take OpenStack. OpenStack is a private cloud. You can deploy OpenStack in your data center and interact with it just like you would with any other cloud service. However, here's the catch – some teams may see it as cloud native and others may see it as general distributed computing. The infrastructure teams that are building out OpenStack will see the *behind-the-scenes* configurations, such as building out the hardware and scaling it across multiple data centers. To them, it's no different than a standard distributed computing environment. Same for the infrastructure engineers that are building, managing, and maintaining the infrastructure for AWS or Azure. Then, there are the teams that interact with OpenStack after it's already built. They're logging into the UI and communicating with OpenStack via the CLI and other API methods, so they're getting the full satisfaction of a true cloud-native environment just like many engineers are getting the full satisfaction of interacting with AWS and Azure without needing to worry about the infrastructure and services on-premises.

Another example is the hybrid cloud. If you're running Azure Stack HCI on-premises, that means you're utilizing some server that runs the Azure Stack HCI operating system, which interacts with the Azure cloud. The engineers that are managing Azure Stack HCI see what's happening behind the scenes. Other engineers that are simply interacting with **Azure Kubernetes Service** (**AKS**) don't see the underlying infrastructure. They just know that they go to a specific location to create a new Kubernetes cluster.

Regardless of where a platform or app is deployed, it could be considered cloud native to some and standard distributed computing to others. You could be an engineer that's building a cloud-native platform so others can interact with it in a cloud-native fashion.

What are microservices?

Taking the idea of distributed computing and cloud native to the next level gives us a microservice. By definition, a microservice is a loosely coupled architecture that has components that have no dependencies on each other.

Say you have five pieces that make up your application: three backend APIs, some middleware to connect the backend and frontend, and a frontend that consists of a website with multiple paths. In a monolithic-style environment, you would take that entire application, package it up, deploy it to a server, and run the binary. Then, if you had to update or upgrade any part of that application, such as the one backend API, you would have to take down the rest of the application. This not only brings down production but also would slow down the ability to get new updates and features out because you'd have to specify a specific window to bring down the entire platform.

Microservices allow you to take those five pieces of the application and split them out into their own individual pieces. Then, you can manage those *pieces* separately instead of having to worry about combining them to deploy and have a working platform.

In a Kubernetes environment, you would have the following:

- One container image for backend API 1
- One container image for backend API 2
- One container image for backend API 3
- One container image for the middleware
- One container image for the frontend

Then, each of those container images can be updated, upgraded, deployed, and managed separately.

It's important to note that microservices aren't just for containers and Kubernetes. The same concepts talked about previously can work just as well on five different Ubuntu VMs. It's just easier to manage a container from a microservice perspective than it is to manage it from a VM perspective. Way less automation and repeatable practices are needed to do the same thing you would have to do on a VM inside of Kubernetes. It's possible and 100% doable, but it takes more effort.

In the next section, you're going to take what you learned in this section and start applying it to Kubernetes-based scenarios.

Learning about Kubernetes app deployments

When engineers are first getting started with deploying an application to a Kubernetes cluster, it looks something like this:

1. Create a Kubernetes manifest.
2. Run a command such as `kubectl apply -f` or `kubectl create -f` against the manifest.
3. Ensure that the application has Pods running.
4. Access your app to ensure it's running the way you were expecting it to run.

Although this is a great approach to getting started with deploying applications to Kubernetes, we must dive a little bit deeper to fully understand how the deployment process of an app occurs, why it works the way that it does, how manifests interact with Kubernetes to ensure an application is deployed, and how Kubernetes keeps the desired state of Pods running.

It seems like how Kubernetes deploys apps is simply magic that occurs on the platform because that's how it's built and that's the way it's supposed to be, but there's a ridiculous number of pieces built that allow Kubernetes to appear to be the magical deployment platform that it makes itself out to be.

In this section, you're going to learn from start to finish how app deployments work inside of Kubernetes and the internals of everything that's needed to make a successful deployment.

Kubernetes manifests

Before actually deploying an application, you'll need to learn the ins and outs of how most applications are deployed to Kubernetes – a **Kubernetes manifest**. The idea is that you already know what a Kubernetes manifest is, but perhaps you don't know the breakdown of the internals of a Kubernetes manifest.

A Kubernetes manifest is a YAML- or JSON-based configuration that interacts differently with the Kubernetes API. The Kubernetes manifest is where you specify what API you want to work with and what resource you want to work with from that API. There are two groups of APIs:

- **Core group**: Consists of the original APIs that Kubernetes came with in `/api/v1`
- **Named group**: Consists of all new APIs that are being built and are in `/apis/$GROUP_NAME/$VERSION`

For example, the following is a code snippet showcasing that the Deployment resource is in the `/api/v1` group:

```
apiVersion: apps/v1
kind: Deployment
```

The following is another example of an Ingress controller, which you can see is in `/apis/networking.k8s.io/v1`:

```
apiVersion: networking.k8s.io/v1
kind: Ingress
```

`apiVersion` is the Kubernetes API you're utilizing, and `kind` is what Kubernetes resource you're creating, updating, or deleting.

A Kubernetes manifest consists of four key parts:

- `apiVersion`: Which version of the Kubernetes API you're using to create the object/resource
- `Kind`: What kind of object you want to create, update, or delete
- `Metadata`: Data that helps uniquely identify the resource/object
- `Spec`: What you want the resource to look like

The following is a Kubernetes manifest:

```
apiVersion: apps/v1
kind: Deployment
```

```
metadata:
  name: nginx-deployment
spec:
  selector:
    matchLabels:
      app: nginxdeployment
  replicas: 2
  template:
    metadata:
      labels:
        app: nginxdeployment
    spec:
      containers:
      - name: nginxdeployment
        image: nginx:latest
        ports:
        - containerPort: 80
```

Let's break that down.

First, you have the API version. You can see that the API version indicates that it's utilizing a resource in the core group. Next, there's kind, which specifies what resource you're creating/updating/deleting. Then there's metadata, which is specifying a name for the deployment to uniquely identify it via metadata. Finally, there's spec, which indicates how you want your containerized app to look. For example, the spec shown earlier indicates that the manifest is using the latest version of the Nginx container image and utilizing port 80.

To wrap up this section, something you should know about Kubernetes manifests, and the way Kubernetes works in general, is that it's declarative. Declarative means *"tell me what to do, not how to do it."* Imperative means *"tell me what to do and how to do it."*

For example, let's say you were teaching someone how to bake a cake. If it was imperative, you would be telling the person what ingredients to use, the size for each ingredient, and how to do it step by step, ultimately leading them to the finished product. Declarative would mean you tell them what ingredients they need and they figure out how to do it on their own.

Kubernetes manifests are declarative because you tell Kubernetes what resource you want to create, including the name of the resource, ports, volumes, and so on, but you don't tell Kubernetes how to make that resource. You simply define what you want, but not how to do it.

The common way but not the only way

Nine times out of ten, when you're deploying a resource to Kubernetes, you'll most likely be using a Kubernetes manifest. However, as you've learned throughout this book so far, the core of Kubernetes is an API. Because it's an API, you can utilize any programmatic approach to interact with it.

For example, the following is a code snippet using Pulumi, a popular IaaS platform to create an Nginx deployment inside of Kubernetes. This code requires more to run it, so don't try to run it. This is just a pseudo example.

There's no YAML and no configuration language. It's raw Go (Golang) code interacting with the Kubernetes API:

```go
        deployment, err := appsv1.NewDeployment(ctx, conf.
Require("deployment"), &appsv1.DeploymentArgs{
        Spec: appsv1.DeploymentSpecArgs{
            Selector: &metav1.LabelSelectorArgs{
                MatchLabels: appLabels,
            },
            Replicas: pulumi.Int(2),
            Template: &corev1.PodTemplateSpecArgs{
                Metadata: &metav1.ObjectMetaArgs{
                    Labels: appLabels,
                },
                Spec: &corev1.PodSpecArgs{
                    Containers: corev1.ContainerArray{
                        corev1.ContainerArgs{
                            Name:   pulumi.String(conf.
Require("containerName")),
                            Image: pulumi.String(conf.
Require("imageName")),
                        }},
                },
            },
        },
    })
    if err != nil {
        return err
    }
```

Although you won't see this too often, you should know that this type of method exists and you can create any resource you want in Kubernetes, in any programmatic fashion, as long as you do it via the Kubernetes API.

Controllers and operators

Kubernetes comes out of the box with a way to ensure that the current state of a deployed application is the desired state. For example, let's say you deploy a Kubernetes deployment that is supposed to have two replicas. Then, for whatever reason, one of the Pods goes away. The deployment controller would see that and do whatever it needs to do to ensure a second replica/Pod gets created. All resources that can be created in Kubernetes (Pods, Services, Ingress, Secrets, and so on) have a **controller**.

An **operator** is a special form of a controller. Operators implement the controller pattern, and their primary job is to move the resources inside of the Kubernetes cluster to the desired state.

Operators also add the Kubernetes API extendibility to use **CustomResourceDefinitions (CRDs)**, which are a way that engineers can utilize the existing Kubernetes API without having to build an entire controller and, instead, just use the CRD controller. You'll see a lot of products/platforms that tie into Kubernetes to create a CRD.

A popular way of building your own operator and controller is with Kubebuilder: `https://book.kubebuilder.io/`.

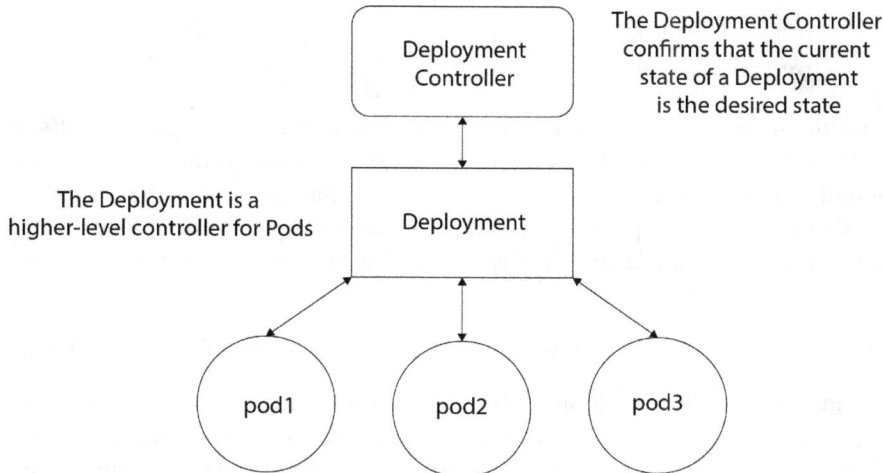

Figure 5.1 – Controllers

You'll hear the terms *operator* and *controller* used interchangeably. To give a frame of reference, just remember that the operator is like the big boss working at a high level, ensuring that things are going well for the organization, and the operator is like the engineer doing the hands-on work to make sure that the organization gets what it needs.

Another form of this that's gaining increased popularity is GitOps. GitOps looks at the desired state of a Kubernetes manifest that's in source control as opposed to what controllers do, which is look at what's actively deployed on a Kubernetes cluster.

Different ways to deploy with higher-level controllers

When you deploy a Pod by itself, the manifest can look like the following:

```
apiVersion: v1
kind: Pod
metadata:
  name: static-web
  labels:
    role: myrole
spec:
  containers:
    - name: web
      image: nginx
      ports:
        - name: web
          containerPort: 80
          protocol: TCP
```

The problem with this method is you have no high-level controller that manages the Pod(s) for you. Without the higher-level controller, like a Deployment or DaemonSet, if the Pod fails, the kubelet watches the static Pod and restarts it if it fails. There's also no management of the current state and desired state. From a production perspective, you never want to deploy a Pod resource by itself. It's fine if you want to test a container image, but that's about it. It should be used for testing/development purposes only.

Not always, but most of the time, you'll see the following production-level controllers that manage Pods:

- **Deployments**: A deployment is one of the highest-level controllers in the Core API group. It gives you the ability to control one Pod or multiple replicas and scale out across the cluster. Deployments also give you the ability to self-heal and confirm that the current state of a deployed containerized application is the desired state. The following code is an example deployment resource:

  ```
  apiVersion: apps/v1
  kind: Deployment
  metadata:
  ```

```
    name: nginx-deployment
spec:
  selector:
    matchLabels:
      app: nginxdeployment
  replicas: 2
  template:
    metadata:
      labels:
        app: nginxdeployment
    spec:
      containers:
      - name: nginxdeployment
        image: nginx:latest
        ports:
        - containerPort: 80
```

- **DaemonSets**: This is like a deployment resource but is cluster wide. It ensures that either all nodes or the nodes you choose run a copy/replica of the Pod. A key difference you'll see in a DaemonSet is that there's no field for replicas. That's because the Pod can't run more replicas than the number of worker nodes, meaning you can't have five Pod replicas if you only have three worker nodes. In that case, you'd only be able to have three Pods. The following code is an example DaemonSet:

```
apiVersion: apps/v1
kind: DaemonSet
metadata:
  name: nginx-deployment
spec:
  # nodeSelecter: Field that you can specify what worker
nodes you want the Pod to deploy to
  selector:
    matchLabels:
      app: nginxdeployment
  template:
    metadata:
      labels:
        app: nginxdeployment
```

```
spec:
  containers:
  - name: nginxdeployment
    image: nginx:latest
    ports:
    - containerPort: 80
```

- **StatefulSets**: This is like a Deployment but for applications that aren't stateless. The StatefulSet maintains a sticky ID for each Pod. For example, let's say that you have an app that needs to communicate with a Pod via a specific network ID or unique ID. With a Deployment, Pods are ephemeral, so they lose their unique ID. With a StatefulSet, a new Pod can get created after the old one failed, but it'll have the same network ID as the old Pod. For a StatefulSet to work, it requires a service to control the network domain. Because its only job is to control the network domain, a headless service makes the most sense. The following is how you can create a standard `StatefulSet`:

```
apiVersion: apps/v1
kind: StatefulSet
metadata:
  name: nginx-deployment
spec:
  selector:
    matchLabels:
      app: nginxdeployment
  serviceName: nginxservice
  replicas: 2
  template:
    metadata:
      labels:
        app: nginxdeployment
    spec:
      containers:
      - name: nginxdeployment
        image: nginx:latest
        ports:
        - containerPort: 80
---
```

```
apiVersion: v1
kind: Service
metadata:
  name: nginxservice
spec:
  selector:
    app: nginxdeployment
  ports:
    - protocol: TCP
      port: 80
  clusterIP: None
```

In terms of which controller to use, it's going to depend on your use case. There's no right or wrong answer here unless it's something straightforward, for example, if you have a containerized app that needs to hold on to its network ID, in which case you'd use a StatefulSet.

Scaling

One of the key components out of the box with Kubernetes is its ability to easily scale both horizontally and vertically. From a production perspective, this takes a ton of load off of your back. In a standard VM environment, you would have to worry about deploying a new server, installing the operating system, getting packages up to date, and deploying the application binary, and finally, the app would be running.

Horizontal Pod autoscaling is the most common, which means more Pods get created to handle load. Vertical autoscaling means the CPU/memory of a Pod gets raised. Vertical Pod autoscaling is not all that common, but possible.

When you're scaling, you can have standard ReplicaSets, but in production, the number may not be so cut and dry. For example, if you have three replicas, but you may need four or ten, you need a way to account for that. The best thing that you can do is start with at least three to four replicas, and if needed, work your way up. If you have to scale up to five or ten, you can update the Kubernetes manifest and redeploy it with a GitOps solution or in another repeatable fashion using the `kubectl apply -f name_of_manifest.yaml` command.

When you're scaling a Pod, or for that matter, when you're doing anything for a Pod deployment, never use commands such as `kubectl patch` or any of the other quick fixes on the command line. If you do, any time the Pod gets redeployed, your configurations won't exist because you did them ad hoc/manually on the command line. Always make changes in a Kubernetes manifest and deploy them properly (remember, current state versus desired state).

How to horizontally scale Pods

When you're scaling Pods horizontally, it's all about replica count. For example, let's say you have a Kubernetes manifest like the following, which contains two replicas:

```yaml
apiVersion: apps/v1
kind: Deployment
metadata:
  name: nginx-deployment
spec:
  selector:
    matchLabels:
      app: nginxdeployment
  replicas: 2
  template:
    metadata:
      labels:
        app: nginxdeployment
    spec:
      containers:
      - name: nginxdeployment
        image: nginx:latest
        ports:
        - containerPort: 80
```

You run `kubectl apply -f nginx.yaml` on the preceding manifest, and then you come to realize that due to user load on the Nginx frontend, you need to bump the replicas from two to four. At that point, you can update the Kubernetes manifest to go from two replicas to four:

```yaml
apiVersion: apps/v1
kind: Deployment
metadata:
  name: nginx-deployment
spec:
  selector:
    matchLabels:
      app: nginxdeployment
  replicas: 4
```

```
template:
  metadata:
    labels:
      app: nginxdeployment
  spec:
    containers:
    - name: nginxdeployment
      image: nginx:latest
      ports:
      - containerPort: 80
```

This method won't recreate anything as you're using `kubectl apply -f` instead of `kubectl create -f`. `create` is for creating net-new resources and `apply` is for updating/patching a resource.

How to vertically scale Pods

Vertically scaling Pods, as discussed, is not a common practice. However, it is doable. The typical method is to use the `VerticalPodAutoscaler` resource from the `autoscaling.k8s.io` API. It gives you the ability to point to an existing deployment so that deployment is managed by the autoscaler. For example, the following Kubernetes manifest shows a target reference of a deployment called `nginxdeployment`:

```
apiVersion: autoscaling.k8s.io/v1
kind: VerticalPodAutoscaler
metadata:
  name: nginx-verticalscale
spec:
  targetRef:
    apiVersion: "apps/v1"
    kind:       Deployment
    name:       nginxdeployment
  updatePolicy:
    updateMode: "Auto"
```

Please note that with the vertical Pod autoscaler turned to `Auto` for the update mode, it'll have the ability to do the following:

- Delete Pods
- Adjust the CPU

- Adjust the memory
- Create a new Pod

It requires a restart of the application running inside the Pod.

Multi-container Pods

Sidecars, sometimes called **multi-container Pods**, are a way to tightly couple containers into one Pod. Typically, and especially from an application perspective, one Pod runs one container. However, there may be use cases where you want to run sidecars. The biggest use case is when you're running some type of log collector/aggregator for your Pods. A lot of engineers will put the container running the log collector into the same Pod where the application is running. That way, it's straightforward to communicate with the application and pull the logs from it as containers inside of a Pod share the same IP address but are reachable on different ports.

One thing you should absolutely never do is run multiple applications in a Pod. For example, you never want to put the frontend app and the backend app inside of the same Pod. That defeats the whole purpose of containers and microservices. Sidecars are only for very specific use cases and if it's absolutely necessary. A personal belief of mine is that you should never use sidecars unless you absolutely have to. Other engineers will disagree with me, but I believe a Pod should run one workload. That's the purpose of a microservice architecture. The only time that I see it absolutely necessary is when you're running a service mesh and you need the service mesh proxy inside of the Pod.

The following is what a Kubernetes manifest would look like if you have multiple containers inside of a Pod. Notice how under `spec.containers`, there's `container1` and `container2`:

```
apiVersion: v1
kind: Pod
metadata:
  name: testsidecar
spec:
  containers:
  - name: container1
    image: nginx
  - name: container2
    image: debian
    command: ["/bin/sh", "-c"]
```

Liveness and readiness probes

Whenever you're deploying any type of application, whether it's containerized or not, you want to ensure that the application is running as expected. Within a containerized environment, it's no different. Let's say you have a Pod that's running an Nginx frontend. The Pod could be up and running, have all of the appropriate resources, and so on. However, that doesn't mean that the binary running inside the Pod is working as expected. To ensure that the actual application is running as expected, you can use **liveness probes** and **readiness probes**.

A liveness probe indicates whether a container is running. It helps Kubernetes understand the overall health of the Pod. The kubelet continuously sends a *ping* of sorts to the container to ensure that it's running as expected. If the liveness probe deems a container unhealthy, the kubelet restarts the Pod.

A readiness probe indicates whether the container is ready to receive requests. Readiness probes are a bit more important from an application perspective because they tell Kubernetes whether or not to route service traffic to Pods. If a service is trying to route traffic to a Pod and that Pod is down or unhealthy, the application won't be reachable. The readiness probe tells the service which Pods are ready to receive requests and which aren't.

The following is an example of a readiness probe:

```
apiVersion: apps/v1
kind: Deployment
metadata:
  name: nginx-deployment
spec:
  selector:
    matchLabels:
      app: nginxdeployment
  replicas: 2
  template:
    metadata:
      labels:
        app: nginxdeployment
    spec:
      containers:
      - name: nginxdeployment
        image: nginx:latest
        imagePullPolicy: Never
        ports:
        - containerPort: 80
```

```
        readinessProbe:
          tcpSocket:
            port: 8080
          initialDelaySeconds: 5
          periodSeconds: 10
```

The following is an example of a liveness probe:

```
apiVersion: apps/v1
kind: Deployment
metadata:
  name: nginx-deployment
spec:
  selector:
    matchLabels:
      app: nginxdeployment
  replicas: 2
  template:
    metadata:
      labels:
        app: nginxdeployment
    spec:
      containers:
      - name: nginxdeployment
        image: nginx:latest
        imagePullPolicy: Never
        ports:
        - containerPort: 80
        livenessProbe:
          httpGet:
            scheme: HTTP
            path: /index.html
            port: 80
          initialDelaySeconds: 5
          periodSeconds: 5
```

All production-level Kubernetes deployments should use readiness probes.

In the next section, you're going to dive into an important topic, which is segregating your containerized apps, and a few different ways of doing it.

Exploring segregation and namespaces

Once applications are deployed, engineers wipe the sweat off their foreheads, give high fives to their team, and rejoice in their victory. However, what comes after the deployment? Better yet, what if you have to deploy the applications again? Or other types of applications? Or to a different location or segregation point? (Segregation will be discussed later in this chapter.) Getting an application up and running is a mental workout in itself, but the *what-comes-next* questions you ask yourself are typically the most important. These are things such as the following:

- Will the next deployment be automated and repeatable?

- If you have to deploy the application again, will it be an effective deployment?

- Can (or should) the apps run right next to each other?

- Which engineers should have access to what apps and why?

Deploying an application is a great victory but designing how and where an application should run is the difference between a successful and a broken-down production environment. Questions around application segregation and multi-tenancy keep engineers up at night because it's less *engineering* work and more planning/architecture work. It's less hands-on-keyboard and more critical thinking at a higher level compared to being down in the trenches in the code.

In this section, you're going to learn about a few of the most popular segregation techniques. Let's get started!

Namespaces

The first level of segregation, typically, is a **namespace**. When you're deploying Pods, the last thing that you want to do is deploy everything and anything to the default namespace. Instead, you want to ensure that applications have their own namespaces. At a network level, Pods within one namespace can communicate with another namespace. However, if you have a service account that's used for Pod deployments in one namespace and a service account that's used to deploy Pods in another namespace, that means the same service account cannot be used to manage all of the Pods. That gives you a bit more segregation from a Pod management perspective and ensures that there's proper authentication and authorization. But from a network perspective, Pods can still communicate with other Pods in separate namespaces.

Notice in *Figure 5.2* that there are three namespaces:

- `argoCD`
- `kube-system`
- `monitoring`

```
argocd       argocd-application-controller-0                   1/1   Running   2 (40h ago)   4d19h
argocd       argocd-applicationset-controller-5b6b596788-9bpnm 1/1   Running   0             4d19h
argocd       argocd-dex-server-7f5957b5df-2xhvb                1/1   Running   0             4d19h
argocd       argocd-notifications-controller-76b9c588c-hwkq6   1/1   Running   0             4d19h
argocd       argocd-redis-ha-haproxy-6dc6757955-65tsn          0/1   Pending   0             4d19h
argocd       argocd-redis-ha-haproxy-6dc6757955-t4str          1/1   Running   2 (40h ago)   4d19h
argocd       argocd-redis-ha-haproxy-6dc6757955-vzrrs          0/1   Pending   0             4d19h
argocd       argocd-redis-ha-server-0                          2/2   Running   2 (39h ago)   4d19h
argocd       argocd-redis-ha-server-1                          0/2   Pending   0             4d19h
argocd       argocd-repo-server-6c75584b95-pg48m               0/1   Pending   0             4d19h
argocd       argocd-repo-server-6c75584b95-sxj2r               1/1   Running   11 (14h ago)  4d19h
argocd       argocd-server-8484cf8cbd-5kbxc                    0/1   Pending   0             4d19h
argocd       argocd-server-8484cf8cbd-hs8k7                    1/1   Running   11 (39h ago)  4d19h
default      nginx-deployment-0                                1/1   Running   0             54m
default      nginx-deployment-1                                1/1   Running   0             54m
default      nginx-deployment-cfmnw                            1/1   Running   0             62m
kube-system  coredns-64897985d-fgcts                           1/1   Running   0             4d19h
kube-system  etcd-minikube                                     1/1   Running   0             4d19h
kube-system  kube-apiserver-minikube                           1/1   Running   0             4d19h
kube-system  kube-controller-manager-minikube                  1/1   Running   0             4d19h
kube-system  kube-proxy-5ttt5                                  1/1   Running   0             4d19h
kube-system  kube-scheduler-minikube                           1/1   Running   0             4d19h
kube-system  storage-provisioner                               1/1   Running   3 (14h ago)   4d19h
monitoring   alertmanager-main-0                               2/2   Running   0             4d19h
monitoring   alertmanager-main-1                               2/2   Running   0             4d19h
monitoring   alertmanager-main-2                               2/2   Running   0             4d19h
monitoring   blackbox-exporter-67bbdf4897-cj4c4                3/3   Running   0             4d19h
monitoring   grafana-86dfcbf9cc-cgs7k                          1/1   Running   0             4d19h
monitoring   kube-state-metrics-7b88fc766c-tzzjv               3/3   Running   0             4d19h
monitoring   node-exporter-b7mZz                               2/2   Running   0             4d19h
monitoring   prometheus-adapter-6455646bdc-6n8zn               1/1   Running   0             4d19h
monitoring   prometheus-adapter-6455646bdc-9s2rr               1/1   Running   0             4d19h
monitoring   prometheus-k8s-0                                  2/2   Running   0             4d19h
monitoring   prometheus-k8s-1                                  2/2   Running   0             4d19h
monitoring   prometheus-operator-5c56ccbbcd-s7v9v              2/2   Running   0             4d19h
```

Figure 5.2 – Kubernetes resources

The preceding screenshot shows that everything in the `argocd` namespace is segregated/isolated from everything in the `kube-system` namespace. If an engineer were to run `kubectl get pods`, they would only see the Pods in the namespace that they have access to.

Single tenancy

Taking segregation and isolation a step further, there are **tenancy models**. First, let's start with **single tenancy**, but before diving in, let's talk about what tenancy models mean in Kubernetes.

Isolating via tenancy could be anything from users to engineers to applications and all different resources. For example, single tenancy could mean running one containerized application across a cluster, or it could mean ensuring that one engineer has access to a cluster that no one else has access to it, but they can run as many applications as they want.

A typical scenario of single tenancy is isolating development environments. Let's say you're a developer and you need a Kubernetes cluster to test an application stack. The scenario would be that the platform engineering team, or whichever team manages Kubernetes clusters, gives you your own Kubernetes cluster to test the application stack. This is a great way to perform single tenancy as it allows all engineers working on different tech stacks to test their code without it compromising or interfering with other application stacks.

Multi-tenancy

On the flip side is **multi-tenancy**. Multi-tenancy is where you have multiple engineers, users, or applications running on the same Kubernetes cluster. If you take a look again at *Figure 5.2* showing the namespaces, you'll see that Prometheus, ArgoCD, and Nginx are running on the same cluster. That would be considered a multi-tenancy cluster. Single tenancy would be if ArgoCD, Nginx, and Prometheus were all running on separate clusters.

In the real world, rarely do you see applications running on different clusters, or rather, an application per cluster. Instead, you usually see the multi-tenancy model for applications and the single-tenancy model for developers testing application stacks, and once the application stack is tested, it moves into the Kubernetes cluster with the rest of the applications.

In the next section, you're going to learn how to think about stateless apps, stateful apps, and volumes.

Investigating stateless and stateful apps

At a high level, applications come in two forms – apps that need data stored and apps that don't care whether the state of the data is stored. Let's think about two different scenarios.

When you log in to your Gmail account, or another email service provider, everything stays where it's supposed to be. You can see the emails in your inbox, the sent messages, the emails in your trash bin, and so on. The application/platform stays how it's supposed to be because the data is stateful. Now, on the opposite side of the spectrum, let's take www.google.com into consideration. When you go to www.google.com in a web browser, you always have a fresh start. The entry box to type in your question is there, but the results to the previous question that you asked Google isn't there. It's always a fresh, clean slate. That's because www.google.com is stateless, as in, it doesn't just hold on to your data (well, it does… but that's a separate discussion) and keep it in the web browser after every search.

Of course, stateless versus stateful is a much deeper discussion, but that's a high-level definition of how you can think about the two different types of applications.

In the next section, you're going to learn about the different deployment methods for stateless and stateful applications inside of Kubernetes, along with resource considerations including limits, quotas, and requests for Pods to ensure that the production-level environment you're running is sustainable.

Stateful versus stateless

In the opening of this section, I shared the Gmail example, which essentially shows what a stateful app is and what a stateless app is. From a Kubernetes perspective, the key difference is that a stateless application doesn't need to store data. Stateful applications require backend storage, such as volumes. Another key difference is that stateful applications require keeping unique IDs, so if a Pod goes down, the Pod that comes up and replaces it must have the same unique ID. A stateless app doesn't need to keep unique IDs.

A common misconception is that stateless apps never use volumes, and that's not the case. You can have a stateless application that, for example, requires a backend database or a volume/hard drive to store values.

Volumes and hard drives aren't what make a stateful application. The unique ID is what makes a stateful application.

Container Storage Interface

For Kubernetes to interact with outside components that aren't native, there must be a plugin of sorts. In the previous chapters, you learned about **Container Network Interface** (**CNI**), which is a plugin to use different network frameworks in Kubernetes. **Container Storage Interface** (**CSI**) is the same thing, but for storage devices. For example, you can have a CSI for NetApp, AWS S3, Azure Storage, and a ton of other storage providers.

Before these interfaces, organizations had to put the code to connect the resources that aren't native in the core Kubernetes code. Just as an example, let's say that Azure wanted to allow Kubernetes engineers to utilize Azure Storage inside of Kubernetes for storing the output of a Pod. Before CSI, Azure would've had to put the code to make it all possible inside of the core Kubernetes code. That was a major hassle because not only did Azure have to wait for a new release of the Kubernetes API to push the feature out, but if there was a bug or a new update that they wanted to push out, they would've had to wait for the next Kubernetes API release.

CSI, and interfaces/plugins across Kubernetes in general, ensures that organizations can create plugins for Kubernetes separately from the core Kubernetes code.

If you want to see an example of CSI, you can check it out on GitHub: `https://github.com/kubernetes-sigs/azuredisk-csi-driver`.

Volumes

Volumes are hard drives, plain and simple.

With a volume, you give a Pod, or multiple Pods, the ability to store data in a location. That location could be Azure, AWS, NetApp, some other storage provider, or even the worker node that the Pod is running on (definitely not recommended. Just an example).

When you're creating a volume for a Pod, there are typically three steps:

- **StorageClass**: A storage class is a way to ask some storage vendor (dynamically) for a hard drive. For example, you can create a storage class that connects to EBS. Then, you can call upon that storage class later with a volume (which you'll learn about in a minute) and utilize the connection to the storage. You can do the same thing in Azure, GCP, and all of the other cloud providers, including most of the storage providers:

```
kind: StorageClass
apiVersion: storage.k8s.io/v1
metadata:
  name: azurefile-csi
provisioner: file.csi.azure.com
allowVolumeExpansion: true
mountOptions:
  - dir_mode=0777
  - file_mode=0777
  - uid=0
  - gid=0
  - mfsymlinks
  - cache=strict
  - actimeo=30
parameters:
  skuName: Premium_LRS
```

- **PersistentVolume**: A persistent volume is created manually by an engineer that uses the storage class to utilize storage from an available source. For example, the Persistent Volume would connect to the EBS storage class from the previous example:

```
apiVersion: v1
kind: PersistentVolume
metadata:
  name: azure-pv
spec:
  storageClassName: " azurefile-csi "
  claimRef:
    name: azurefile
    namespace: default
```

- **PersistentVolumeClaim**: The last piece is the persistent volume claim, which is a request made by a user, usually in a Kubernetes manifest that's creating a Pod, to use some of the storage that's available in the storage class. The engineer can say "hey, I want 10 GB of storage from this storage class for my Pod":

```
apiVersion: v1
kind: PersistentVolumeClaim
metadata:
  name: azurefile
spec:
  accessModes:
    - ReadWriteMany
  storageClassName: azurefile-csi
  resources:
    requests:
      storage: 10Gi
```

At this point, you may be wondering "well, why do I need a persistent volume if I can just automatically request some storage with a claim?", and that's a good question. The answer is going to depend on your environment. If you're using NetApp storage, and you have 1,000 GB of storage, you want an engineer to create a persistent volume and manage those volumes because you only have 1,000 GB of storage. If you attempt to go over that 1,000 GB, failures will start to occur, so having someone manage it makes sense. On the flip side, if you're using cloud storage, such as in Azure or AWS, that storage is *unlimited* to a user (you of course have to pay for it), so going straight to a persistent volume claim instead of having an engineer create a persistent volume would make sense.

Resource requests and limits

In any production environment, you have Kubernetes clusters that are running on servers, regardless of whether it's on-premises or a managed Kubernetes service. Because they are running on servers, those servers have hardware resources, and all servers have a limit. There's no *unlimited CPU* on a server or *unlimited memory*. There are limits to a server's resources and servers can reach 100% capacity.

Because of that, when you're creating Pods, you should specify limits and requests. You never want to give anything, whether it's a virtualized VM or a containerized app, open reign in an environment to take as much CPU and memory as it wants. If you don't control resources, such as memory and CPU, every application could take whatever resources it wanted to take.

Let's think about a basic example. Say you have an application that has a memory leak. If you containerize it, the Pod that it's running in will continue to take more and more memory until the worker node eventually fails and/or the application crashes, and you'll only know when it's too late.

Before diving in, let's define the difference between a **limit** and a **request**.

A limit is telling a Pod "you cannot go above this." For example, if you specify X amount of CPU or memory on a Pod, that Pod cannot go above that limit. It's completely blocked.

The following is an example of a limit. As you can see, the Nginx app is limited to 128 Mi of memory. Anything above that and Kubernetes will say "nope, you can't have it":

```
spec:
  containers:
  - name: nginxdeployment
    image: nginx:latest
    resources:
      limits:
        memory: "128Mi"
    ports:
    - containerPort: 80
```

A request is what the Pod is guaranteed to get. If a Pod requests a resource, Kubernetes will only schedule it on a worker node that can give it that resource.

The following is an example of a request. In this example, Kubernetes will say "alright, you want 64 Mi of memory and 250m of CPU. Let me schedule you onto a worker node that can handle this":

```
spec:
  containers:
  - name: nginxdeployment
    image: nginx:latest
    resources:
      requests:
        memory: "64Mi"
        cpu: "250m"
    ports:
    - containerPort: 80
```

The following is an entire manifest example:

```
apiVersion: apps/v1
kind: Deployment
metadata:
  name: nginx-deployment
```

```
spec:
  selector:
    matchLabels:
      app: nginxdeployment
  replicas: 2
  template:
    metadata:
      namespace: webapp
      labels:
        app: nginxdeployment
    spec:
      containers:
      - name: nginxdeployment
        image: nginx:latest
        resources:
          requests:
            memory: "64Mi"
            cpu: "250m"
          limits:
            memory: "128Mi"
        ports:
        - containerPort: 80
```

Which should you choose?

There's some confusion around how requests and limits work.

When Pods are done using memory, they give that memory back to the worker node and it goes back into the pool for other Pods to use. With the CPU, it does not. The Pod will hold on to that CPU. Because of that, it's not a best practice to let the Pod just hold on to the CPU until it gets deleted because it may not always need that amount of CPU. It's essentially wasting CPU resources.

So, which should you choose?

In every production environment, you should always set up requests, but you should only limit CPU.

Namespace quotas

When it comes to limits and requests, one of the really awesome things that you can do is set them up for namespaces. For example, you can have a namespace that has a limit of 1,000 Mi and a request of 512 Mi. That way, all nodes running in that namespace automatically get limited to the required resources, which means you don't have to put limits and requests into every single Kubernetes Pod manifest. The following code block showcases the resource quota:

```
apiVersion: v1
kind: namespace
metadata:
  name: test
---
apiVersion: v1
kind: ResourceQuota
metadata:
  name: memorylimit
  namespace: test
spec:
  hard:
    requests.memory: 512Mi
    limits.memory: 1000Mi
```

In the next and final section, you're going to learn how to upgrade apps and different types of update methods.

Upgrading Kubernetes apps

Throughout this chapter, you learned some very important lessons:

- How to deploy an app
- How to deploy different types of apps on Kubernetes
- How to ensure apps are properly scaled
- How to ensure apps are running as you expected

Once you get an application to where you'd like it to be, it's a great accomplishment. Then, before you know it, it's time to upgrade or update the application and you have to start on the journey all over again. You must test out the new version of the app, get it deployed without taking down the entire production environment, and retest all the components to ensure it's running as expected.

There may also be times, which is extremely common, when you must roll back an update or upgrade to a previous application version. Perhaps it wasn't properly tested in the staging environment, or something popped up that the QA/regression testing didn't catch. In any case, you need a solid plan and methodology on how to do a rollback.

In this section, you're going to learn a few different ways to test out application updates and upgrades, how you can upgrade and update applications running in Kubernetes, and how you can roll back updates and upgrades when necessary.

Types of upgrades

First, let's break down the typical types of upgrades in Kubernetes.

A/B testing is a way to have a set of users on one version of the application and a set of users on another version of the application. For example, let's say you're testing out two versions of an app, v1.1 and v1.2. A set of users would get v1.1 and another set of users would get v1.2. At that point, you can test things such as performance, how the users are interacting with the new version of the app, bugs, and issues. This type of test is a controlled experiment.

Canary deployments are pretty much identical to A/B testing except they're done with real users. Taking the previous example, let's say you had v1.1 and v1.2 of an app. You would roll out v1.2 in production and put a set of users on v1.2 but keep a set of users on v1.1. That way, you can see how users interact with the new version in production.

Blue/green testing is when you have two production environments, one on v1.1 and one on v1.2. All the users are still on v1.1, but you slowly start to migrate all of the users to v1.2. All users are moved over to v1.2 once it's confirmed to be working.

In Kubernetes, the most popular upgrade method is a **rolling update**, which, based on the preceding explanations, is a blue/green deployment.

What happens to an app being upgraded?

When you're upgrading a container image in a Pod, what happens is the new Pod comes up and is tested and the old Pod then gets deleted.

Let's take the example from the previous section regarding v1.1 and v.1.2 with the help of the following diagram:

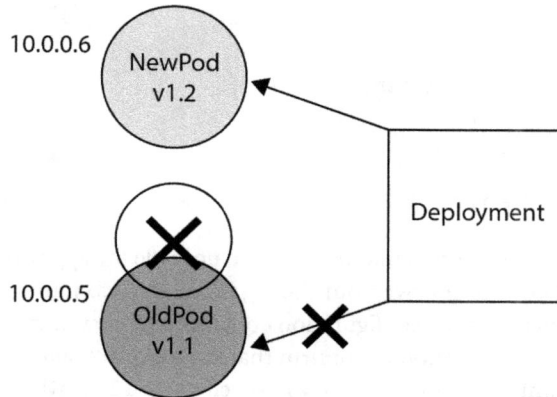

Figure 5.3 – Rolling update

In the preceding architecture diagram, what's happening is v1.1 is running on a Pod with an IP address of 10.0.0.5. Then, the new Pod running v1.2 comes up and is running at the same time as the old Pod. Once the deployment confirms that v1.2 of the Pod is working properly and as expected, the users will begin to move over to the new Pod. Once all users are on the new Pod running v1.2, the old Pod running v1.1 gets deleted.

Rolling updates

What was explained in the previous section was a rolling update. Let's take a look at it from a code perspective. The following is a Kubernetes manifest that's running a deployment spec with a containerized Nginx image using v1.1:

```
apiVersion: apps/v1
kind: Deployment
metadata:
  name: nginx-deployment
spec:
  selector:
    matchLabels:
      app: nginxdeployment
  replicas: 2
  template:
    metadata:
      labels:
        app: nginxdeployment
```

```
spec:
  containers:
  - name: nginxdeployment
    image: nginx:1.1
    ports:
    - containerPort: 80
```

Then, the time comes to upgrade the containerized app. To upgrade the app with `RollingUpdate` (blue/green deployment), you would swap out the `nginx:1.1` container image version with `nginx:1.2`. The `RollingUpdate` configuration contains a `progressDeadlineSeconds` and `minReadySeconds` configuration to confirm that the new version of the containerized app comes up appropriately. Within the strategy map, you specify a `RollingUpdate` type and ensure that one replica is always running the old containerized app version as the update occurs. That way, users aren't kicked off the app. The following code will perform the proper rolling update action:

```
apiVersion: apps/v1
kind: Deployment
metadata:
  name: nginx-deployment
spec:
  selector:
    matchLabels:
      app: nginxdeployment
  revisionHistoryLimit: 3
  progressDeadlineSeconds: 300
  minReadySeconds: 10
  strategy:
    type: RollingUpdate
    rollingUpdate:
      maxUnavailable: 1
      maxSurge: 1
  replicas: 4
  template:
    metadata:
      labels:
        app: nginxdeployment
    spec:
      containers:
```

```
    - name: nginxdeployment
      image: nginx:1.2
      ports:
      - containerPort: 80
```

You would then run `kubectl apply -f` against the Kubernetes manifest, and the rolling update would begin.

Rollbacks

If you'd like to roll back `RollingUpdate`, you'll need two commands.

First, get the revision number that you want to roll back to from the following command:

```
kubectl rollout history deployment nginxdeployment
```

Next, undo `RollingUpdate`:

```
kubectl rollout undo deployment/nginxdeployment
--to-revision=whichever_revision_number_youd_like
```

Not only are updates and rollbacks important to understand from an educational perspective, but you'll most likely see this a fair amount as your organization moves to a more microservice-driven approach.

Summary

There are many types of resource deployments when it comes to Kubernetes and often, there's no right or wrong answer to which you choose. The only time that there's a true right or wrong answer is depending on the deployment. If you have a stateful application, you want to use a StatefulSet. There's no mystery as to which controller you should be using and there's no *good or bad*. It simply depends on the type of application and workload you need to deploy and manage.

In the next chapter, we'll be diving a little bit deeper into different types of deployments from a more advanced perspective.

Further reading

- *Kubernetes – An Enterprise Guide* by Marc Boorshtein and Scott Surovich: `https://www.packtpub.com/product/kubernetes-an-enterprise-guide-second-edition/9781803230030`

6

Kubernetes Deployment– Same Game, Next Level

In the previous chapter, you dove into different deployment scenarios and how you should think not only about controllers, but also about upgrading apps, different types of apps to deploy, and different methods for getting an app up and running. In this chapter, you're going to dive a bit deeper into the different styles of deploying and troubleshooting versus just doing the deployment.

The first step in any type of deployment is figuring out what you're doing – what type of application it is, what type of Kubernetes resource you want to use, and the different plugins that you may want to use, such as a CSI. After you figure out the logistics of what you want to deploy, the next step is to think about how you want to deploy.

With Kubernetes, there are many different deployment methods – automated deployments, manual deployments, and something in between automated and manual. There's a vast number of different ways to perform deployments, so you won't learn about all of them because that could take up more than six chapters of a book in itself, but you will learn about the primary ways to deploy and package up Kubernetes Manifests.

After you learn about deployments, thinking about how to troubleshoot once something inevitably goes wrong is a good, logical next step. Typically, engineers will learn troubleshooting on the fly, but it's a good approach to think about troubleshooting techniques prior to something going wrong.

After learning about troubleshooting and deploying containerized apps, you're going to wrap up with how to manage network connectivity between apps running on Kubernetes and how to migrate existing, more monolithic-style applications.

In this chapter, we're going to cover the following topics:

- Getting to know Helm charts and Kustomize
- Deploying with CI/CD and GitOps
- Troubleshooting application deployments
- Service meshes and Ingresses

Technical requirements

In this section, you're going to take what you learned about the different types of deployments and methodologies for thinking about Kubernetes resources from the last chapter and expand upon that knowledge in this chapter. You should have a brief understanding of automated deployment methodologies such as CI/CD, and have a high-level understanding of what a service mesh is, along with some application architecture knowledge. As usual, you'll find the code for this chapter on GitHub: `https://github.com/PacktPublishing/50-Kubernetes-Concepts-Every-DevOps-Engineer-Should-Know/tree/main/Ch6`.

Getting to know Helm charts and Kustomize

When you're working with Kubernetes, unless it's a dev environment for your testing, there's an extremely slim chance that you only have one Kubernetes Manifest. You most likely have multiple for various resources such as Deployments, Services, DaemonSets, ConfigMaps, Ingresses, and a ton of the other Kubernetes resources out there. Utilizing almost every single Kubernetes platform or tool that's deployed to your cluster uses a Kubernetes Manifest.

With all those Kubernetes Manifests, there are a ton of different values and parameters that you need to pass at runtime to make it all work. In this section, you'll learn about two different methods of managing Kubernetes Manifests – **Helm charts** and **Kustomize**.

Why think about deployment methods for manifests?

Before diving into different deployment methods, it makes sense to understand why you'd want to consider deployment methods other than using the terminal for the deploying Manifests first.

There are three primary points, which we discuss in the following subsections.

Scale

When thinking about scale, there's absolutely no way to scale a deployment if an engineer is always doing it from their laptop. The engineer could be using different plugins, different IDEs, different terminal settings, and even a different operating system. With all of that, the uncertainty alone of the environment can cause a massive amount of error. If every engineer is relying on their computer to deploy an environment, what happens if their laptop crashes? Or there's a random update during the day? Or someone is out of office? There are so many variables that come into play that make utilizing a local computer a bad idea when it comes to deployments. Instead, it makes far more sense to have a central location from which you conduct your deployments. The environment stays the same, everyone can use it, you can customize it to your team's needs, and you don't have to worry about anyone being out of office.

Anything can go wrong

Going into the second point, which echoes the first point in a sense, anything can go wrong. The goal of every organization is to have a successful deployment all the time, zero hiccups, and the ability to deploy at any time. Marketing teams paint this picture in our heads of "deploy 20 times per day with this tool and it'll always work," but as all engineers know, that's not reality. Anything as simple as a network hiccup or making an error when entering a variable name can lead to a failed deployment, and, in turn, an application being down. Because of that, having a proper deployment strategy is key not only to repeatability with Kubernetes but also to repeatability in general. Having a proper process and *rules* in place of how something is deployed and when or where it's deployed is the make or break between a successful update and everyone on the engineering team sitting in the office fixing an issue until 1:00 A.M.

It's manual

The last point, which goes without saying, is that it's an incredibly manual process to sit at a terminal and run commands to deploy a configuration. In today's world, engineers want to spend their time focusing on value-driven work, not putting out fires. In fact, that's a huge reason why automation and repeatability exist in the first place. Engineers wanted to get their time back and stop working on mundane tasks. If you're constantly deploying on your computer to an environment, you're putting the "this is awful" back into manual work. Now, there are circumstances where you'd want to deploy from your localhost. For example, when I'm deploying to a dev environment or testing a new config for the first time, I'm not going to create a repeatable solution around it because I'm unsure whether it even works yet. However, once I know that it works and my initial dev testing is complete, I'll automate the workflow.

Going forward in this chapter, keep in mind that the reason why you want to think about deployment workflows is to mitigate as much of the three aforementioned points as possible.

Helm charts

The idea behind repeatable deployment strategies is to make your life easier, but with new strategies comes the need to learn about different methods of implementation. The first method to learn about is Helm charts.

Helm is an open source project originally created by DeisLabs and donated to the CNCF; the CNCF now maintains the project. The objective of Helm when it first came out was to provide engineers with a better way to manage all the Kubernetes Manifests created. Helm was built with Kubernetes in mind and it's a tool and platform specifically for Kubernetes, so it's the same YAML you're used to, just packaged differently – literally just YAML. Kubernetes was meant to give you a way to declaratively deploy containerized apps. It wasn't necessarily meant to give you a meaningful way to package a bunch of Kubernetes Manifests so you could use them together. That's where Helm comes into play. In addition, Helm keeps a release history of all deployed charts. This means you could go back to a previous release if something went wrong.

In January 2016, the project merged with a GCS tool called Kubernetes Deployment Manager, and the project was moved under Kubernetes. Helm was promoted from a Kubernetes subproject to a CNCF project in June 2018.

In short, Helm is a way to take a bunch of Kubernetes Manifests and package them up to be deployed like an application.

Using Helm charts

Now that you know about Helm, let's go ahead and dive into it from a hands-on perspective:

1. The first thing that you'll need to do is install Helm. Because it varies based on the operating system, you can find a few different installation methods here: `https://helm.sh/docs/intro/install/`.

2. Once Helm is installed, find or create a directory in which you want your first Helm chart to exist. Preferably, this will be an empty directory:

    ```
    mkdir myfirsthelmchart
    ```

3. Next, go into that directory on your terminal.

4. In the new directory, run the following command to create a Helm chart:

    ```
    helm create name_of_chart
    ```

    Once you do that, you should see a directory structure similar to the following screenshot. In this case, the chart was called `newchart`.

Figure 6.1 – Helm chart

If you open up the `templates` directory, you'll see a bunch of examples for Deployments, Ingresses, and a lot more.

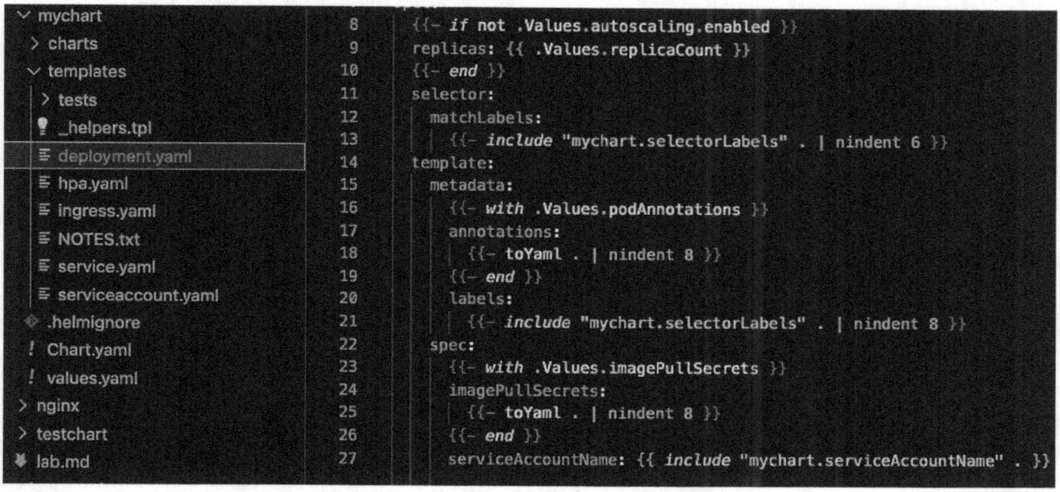

Figure 6.2 – Example Helm

If you open up `values.yaml`, you'll see where you can start adding values that you want to pass into your templates.

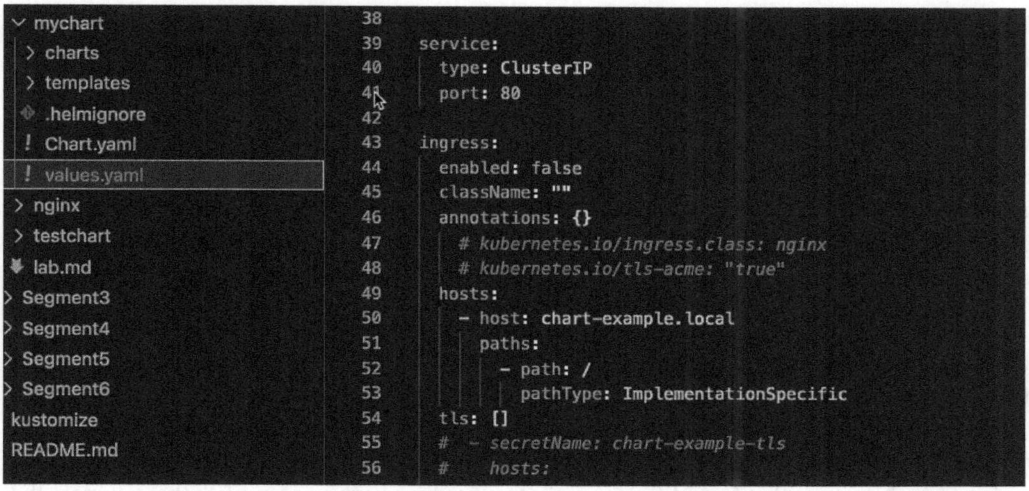

Figure 6.3 – Values file

5. To deploy a Helm chart, run the following command:

```
helm install nginxapp .
```

6. To install the Helm chart, run the following command. The . symbol indicates the current directory, which is where the Helm chart exists:

```
helm install mynewapp .
```

Of course, this isn't everything there is to know about Helm. In fact, there are literally entire books on Helm. The goal of this section was to get you on the right path.

Helm chart best practices

The following is a list of best practices to follow in production when using Helm:

- When storing Helm charts, ensure that they're set to be public or private as required. The last thing you want is to push a Helm chart to a registry that's public-facing when it's not supposed to be.

- Document what your charts do.

- Ensure you store charts in source control.

- Always test Helm charts after a change is made.

Kustomize

Helm and Kustomize are pretty similar but have some unique differences. One of the primary use cases of Helm is to have a values.yaml file to store values to pass into a Kubernetes Manifest. Kustomize has the same type of idea.

With Kustomize, you have a template, typically called a base. The base is the template that you want to use. It could be for a Kubernetes Deployment, Service, Pod, or anything else you'd like. The template is the literal base where your values get pushed into. Along with the template, you have a kustomization.yaml file, which tells Kustomize which templates to use. For example, let's say you have a deployment.yaml and service.yaml file. You would put those two filenames into the kustomization.yaml file so Kustomize knows it should push values into those two files.

Values were mentioned a few times already, but not thoroughly explained. A value can be anything that you want to essentially pass in at runtime. For example, let's say you have three environments – dev, staging, and prod. In dev, you have one replica. In staging, you have two replicas. In prod, you have three to four replicas. You can use Kustomize to pass those values into one template, so instead of having three Manifests that have different replica values, you have one template that you pass the values into.

But how do you pass in the values?

Within a Kustomize directory, you typically have two directories – base and overlays. The base is where the template goes. The overlay directory is where each environment goes with specific values. For example, let's say you have a dev, staging, and prod overlay.

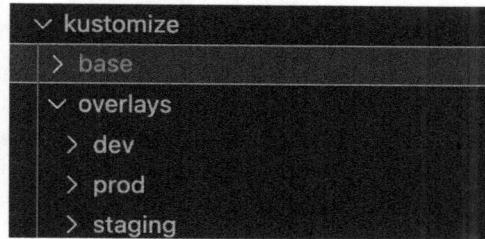

Figure 6.4 – Base configuration

The dev overlay, along with the others, would have a kustomization.yaml file.

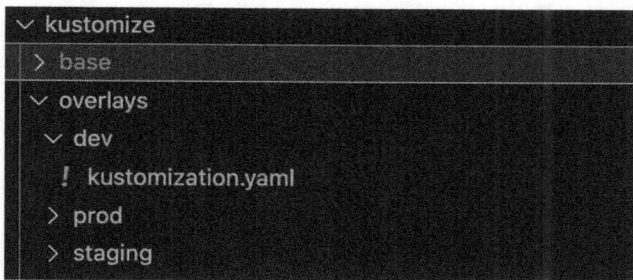

Figure 6.5 – Dev overlay configuration

Within the kustomization.yaml file is where you'd find the config for the replica count.

```
apiVersion: kustomize.config.k8s.io/v1beta1
kind: Kustomization
resources:
- ../../base/

replicas:
- name: nginx-deployment
  count: 1          You, 4 months ago • kustomize
```

Figure 6.6 – Kustomization file

Notice how the resources map is pointing to the base directory, and the replicas map is specifying the deployment along with the replica count.

The primary difference between Helm and Kustomize is that Helm's primary purpose is to package up a bunch of Kubernetes Manifests and deploy them like an app, whereas the primary purpose of Kustomize is to have a template that you push your values into (such as the replica count). Helm does this as well, but it's not the primary purpose of Helm.

Using Kustomize configurations

Now that you know about Kustomize, let's dive into it from a hands-on perspective:

1. The first thing that you'll need to do is install Kustomize. Because it varies based on the operating system, you can find a few different installation methods here: `https://kubectl.docs.kubernetes.io/installation/kustomize/`.

2. Once Kustomize is installed, find or create a new directory in which you want your Kustomize config to live. You can call it `kustomize`.

3. Create two new directories under the `kustomize` directory called `overlays` and `base`. Inside the `overlays` directory, create a new child directory called `dev`. It should look similar to the following screenshot.

Figure 6.7 – Dev overlay configuration

4. Inside the `base` directory, create a new file called `deployment.yaml` and paste the following code into it:

```
apiVersion: apps/v1
kind: Deployment
metadata:
  name: nginx-deployment
spec:
  selector:
    matchLabels:
      app: nginxdeployment
  replicas: 2
  template:
    metadata:
      labels:
        app: nginxdeployment
    spec:
      containers:
      - name: nginxdeployment
        image: nginx:latest
        ports:
        - containerPort: 80
```

5. Next, create a new file in the `base` directory called `kustomization.yaml` and paste the following configuration into it, which tells Kustomize which Kubernetes Manifest to utilize:

```
apiVersion: kustomize.config.k8s.io/v1beta1
kind: Kustomization
resources:
  - deployment.yaml
```

6. For the last step, inside of `overlays | dev`, create a new file, call it `kustomization.yaml`, and paste the following Manifest into it:

```
apiVersion: kustomize.config.k8s.io/v1beta1
kind: Kustomization
resources:
- ../../base/

replicas:
- name: nginx-deployment
  count: 1
```

7. Once the directories and configurations are in place, `cd` into the `base | dev` directory and run the following command:

```
kubectl kustomize
```

You'll see an output similar to the following screenshot, which gives you a config with one replica, instead of two, which is what the template contains.

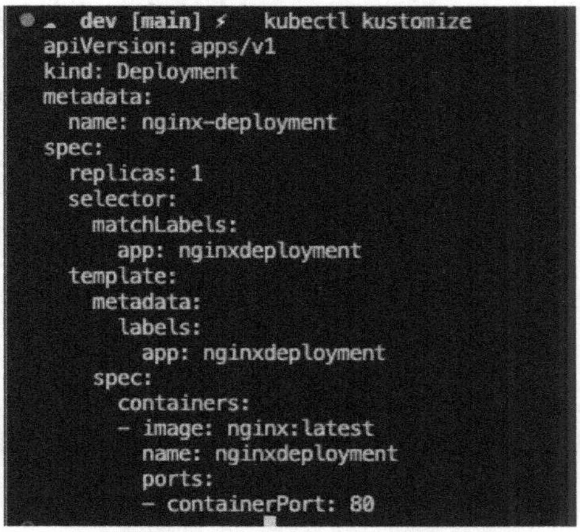

Figure 6.8 – Kustomize output

As with Helm charts, the topic of Kustomize could fill a small book itself, which means this section couldn't cover it all. It should, however, get you started in the right direction.

Kustomize best practices

The following is a list of best practices to follow in production when using Kustomize:

- Ensure that you put overlays into their own directories. You don't have to do this, but it makes for a much cleaner config.

- Ensure that all code is stored in source control.

- Follow a standard directory structure – `base` for the directory where the template goes and `overlays` for values that you wish to pass into the template.

In the next section, you'll learn about the two primary deployment methods when it comes to containerized apps.

Deploying with CI/CD and GitOps

Kubernetes deployments come in three stages:

- Deploying Manifests on your local computer

- Deploying Manifests with an automated solution such as CI/CD, which ultimately just runs `kubectl apply -f` commands, the same as your local computer

- A new and completely automated solution that's (usually) 100% hands-off from a deployment perspective

With the first stage, it wasn't scalable at all. A bunch of engineers were running commands on their localhost to deploy a containerized app, and they were all doing it in different ways with different code editors and different plugins. It was a mess and didn't allow scalability for the deployment process. It also held engineers up from doing value-driven work and instead, they had to sit on their terminals and run commands all day.

In this section, you'll learn about more common, automated, and new approaches to deploying apps, which will be around CI/CD and GitOps.

What is CI/CD?

When it comes to CI/CD, it's assumed that if you're reading this book, you're already doing work in CI/CD and know what it is. Because of that, there won't be an entire breakdown of CI/CD. Let's do a brief overview.

By definition, CI/CD is a way to create an artifact of your application and deploy it to the desired destination in an automated fashion. As CI/CD increased in popularity, engineers started using it for other purposes – for example, packaging up Terraform code into an artifact and running it so infrastructure can be created automatically.

In the CI process, engineers are worried about the following:

- Testing code

- Packaging up code

- Ensuring that all prerequisites and dependencies are met

- Building container images

In the CD process, engineers are worried about the following:

- Deploying workloads

- Ensuring that they reached the correct destination

- Ensuring that the app or services and infrastructure are up and running as expected

Using CI/CD for Kubernetes deployments

As with everything else in tech, there are what feels like a million ways to do one thing. Because of that, we cannot specify every CI/CD, automation, and cloud scenario here. To make things simplistic, Terraform code for GKE and YAML pipelines for GitHub Actions will be shown. This is considered pseudocode, but it'll actually work in the right environments.

First, let's start with the Terraform code and break it down:

1. You'll start with the Google provider, specifying the region:

    ```
    provider "google" {
      project    = var.project_id
      region     = var.region
    }
    ```

2. Next, `google_container_cluster` will be specified so you can implement the VPC you want to use, subnet, and worker node count:

    ```
    resource "google_container_cluster" "primary" {
      name     = var.cluster_name
      location = var.region
    ```

```
  remove_default_node_pool = true
  initial_node_count       = 1

  network    = var.vpc_name
  subnetwork = var.subnet_name
}
```

3. The last resource is for `google_container_node_pool`, which implements the needed Google APIs for GKE, the node count, node names, and node size or type:

```
resource "google_container_node_pool" "nodes" {
  name       = "${google_container_cluster.primary.name}-node-pool"
  location   = var.region
  cluster    = google_container_cluster.primary.name
  node_count = var.node_count

  node_config {
    oauth_scopes = [
      "https://www.googleapis.com/auth/logging.write",
      "https://www.googleapis.com/auth/monitoring",
    ]

    labels = {
      env = var.project_id
    }

    machine_type = "n1-standard-1"
    tags         = ["gke-node", "${var.project_id}-gke"]
    metadata = {
      disable-legacy-endpoints = "true"
    }
  }
}
```

With the Terraform code, you'll want a way to deploy it. The best way in today's world is with CI/CD. When it comes to deploying infrastructure and services, CI/CD is a great and repeatable process.

To deploy the code, you can use any CI/CD platform of your choosing, but the code here is an example of how you can deploy the Terraform code via GitHub Actions.

The pipeline does the following:

- Specifies `workflow_dispatch`, which means the code will only run if you click the `Deployment` button

- Uses an Ubuntu container to run the pipeline

- Checks out the code (clones it) to the Ubuntu container

- Configures Terraform in the Ubuntu container

- Configures the GCP SDK in the Ubuntu container

- Runs `terraform init`, and formats, plans, and applies it to the directory where the GKE code lives:

```
name: GKE Kubernetes Deployment
on:
  workflow_dispatch:

jobs:
  build:
    runs-on: ubuntu-latest

    steps:
      - uses: actions/checkout@v3

      - name: Setup Terraform
        uses: hashicorp/setup-terraform@v1

      - name: Set up gcloud Cloud SDK environment
        uses: google-github-actions/setup-gcloud@v0.6.0
        with:
          service_account_email:
          service_account_key:
          project_id:

      - name: Terraform Init
```

```
            working-directory: where_the_gke_code_lives
            run: terraform init

        - name: Terraform Format
            working-directory: where_the_gke_code_lives
            run: terraform fmt

        - name: Terraform Plan
            working-directory: where_the_gke_code_lives
            run: terraform plan

        - name: Terraform Apply
            working-directory: where_the_gke_code_lives
            run: terraform apply -auto-approve
```

When using CI/CD, it makes the most sense to use it in this type of way for Kubernetes. You rarely ever want to use CI/CD to deploy a Kubernetes Manifest and instead, you'd want to use something such as a GitOps solution, as it's far more efficient, manages the state, monitors the workloads, and a lot more.

What is GitOps?

By definition, GitOps is a set of tools that utilizes Git repositories as a source of truth to deliver Kubernetes resources as code. It's an operational best practice used for app development, collaboration, compliance, and CI/CD, and applies the best practices to infrastructure automation. Now, let's see a simpler explanation. It's configuration management for Kubernetes; that's it, plain and simple. Configuration management is all about ensuring that the desired state is the current state, which is what GitOps gives us.

Now that you know the definition of GitOps, let's talk about what it actually does for Kubernetes. First things first, you have a source control repository. The repository contains your Kubernetes manifests that you wish to deploy to Kubernetes to run your applications. You also have a Kubernetes cluster, which is running on any environment you'd like. It could be on-premises, in a raw Kubernetes cluster, or even in a cloud-based service such as GKE or EKS. Now, you have the Kubernetes Manifests that you want to run in your production environment and the Kubernetes cluster that you want to run the Kubernetes manifests on, but how do you deploy them? The typical way is running something such as `kubectl apply -f` against the Kubernetes manifests, but that requires manual effort and leaves a lot to be desired. Instead, you can implement GitOps. To implement GitOps, there are a few solutions. You decide to implement a GitOps solution, and that GitOps solution needs access to both the Kubernetes cluster that you're running and the source control system, such as GitHub or any other Git system where your source code is stored. To do that, you install the GitOps solution on the Kubernetes cluster and while doing that, you give the GitOps solution access to your source control system with some type of **Personal Access Token** (**PAT**) or another type of authentication or authorization method. After that, you use the GitOps solution to deploy the Kubernetes Manifests that live in source control. At this stage, you're not using `kubectl apply -f` or `kubectl create -f` anymore. Instead, you're using the CLI or whatever other solution comes with the GitOps platform to deploy the Kubernetes Manifests – and boom, just like that, you have an application deployed! Now, of course, we all wish it were that easy. A couple of sentences to explain and poof, you're up and running in a production environment. However, it's not that simple, which is why GitOps is in such high demand and isn't the easiest thing to crack.

At the time of writing this, the most popular GitOps platforms are **ArgoCD** and Flux.

Using GitOps for automated deployments

Knowing the process to create the Kubernetes infrastructure, you can now deploy and manage a containerized app using GitOps. To follow this section, you'll need a Kubernetes environment up and running, as ArgoCD will be deployed to the cluster.

This section is going to be more of a step-by-step guide because regardless of where you're running Kubernetes, these are the steps to get ArgoCD up and running. Unlike with the CI/CD section, there aren't tons of different platforms, cloud environments, or configuration code choices that can come into play, and because of that, the following solution can work in any environment.

Configuring ArgoCD

1. First, create a namespace for ArgoCD in your Kubernetes cluster:

   ```
   kubectl create namespace argocd
   ```

2. Install ArgoCD using the preconfigured Kubernetes Manifest from ArgoCD that provides a highly available installation:

   ```
   kubectl apply -n argocd -f https://raw.githubusercontent.
   com/argoproj/argo-cd/stable/manifests/ha/install.yaml
   ```

Figure 6.9 – ArgoCD creation output

3.   Get the initial admin password to log in to ArgoCD:

```
kubectl get secret -n argocd argocd-initial-admin-secret
-o jsonpath="{.data.password}" | base64 -d
```

4.   Open up ArgoCD's UI via Kubernetes port forwarding. That way, you can access the frontend of ArgoCD without having to attach a load balancer to the service:

```
kubectl port-forward -n argocd service/argocd-server :80
```

5.   Now that you know the UI works, log in to the server via the CLI. That way, you can deploy containerized apps with ArgoCD via the CLI to create a repeatable process instead of doing it through the UI, which is manual and repetitive.

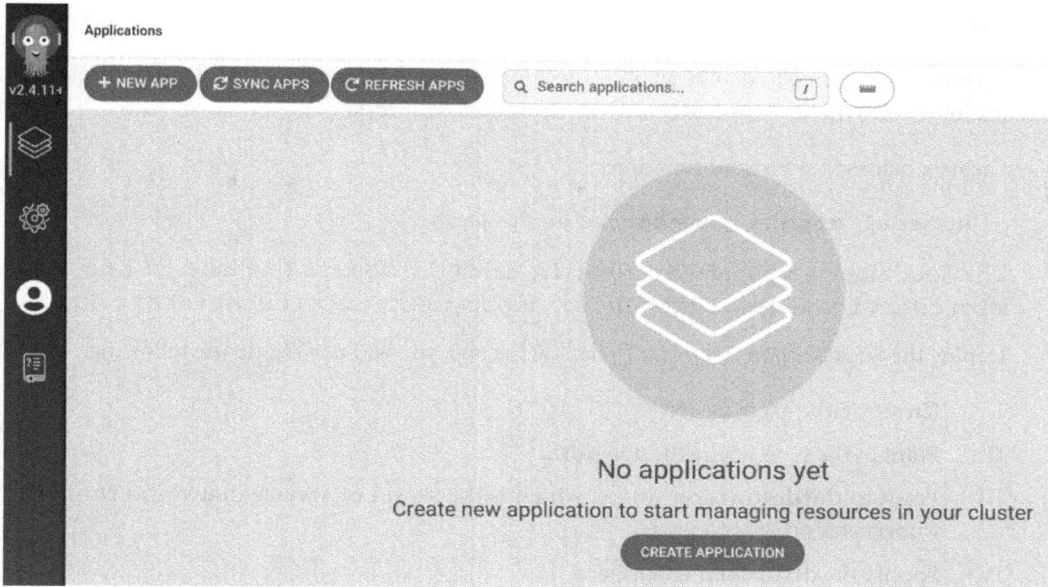

Figure 6.10 – ArgoCD portal

6. The port is what ArgoCD is hosting from the `kubectl port-forward` command that you ran in the previous step. Use the following command to log in to ArgoCD:

```
argocd login 127.0.0.1:argocd_port_here
```

```
WARNING: server certificate had error: x509: "Argo CD" certificate is not trusted. Proceed insecurely (y/n)? y
Username: admin
Password:
'admin:login' logged in successfully
Context '127.0.0.1:59341' updated
```

Figure 6.11 – Login output

7. In the Argo CD UI, go to **User Info | Update Password**. Change the password from the initial admin password to a password of your choosing.

You now have officially deployed ArgoCD and have the ability to work with the GitOps platform on your terminal and locally on your computer.

Deploying an app

In this section, you're going to deploy an app. The app that you'll use is a very popular demo-related app that a lot of folks use to showcase how an environment will work:

1. Create a namespace for your new app:

    ```
    kubectl create namespace sock-shop
    ```

    The Sock Shop is a popular microservice demo that you can find here: https://microservices-demo.github.io/deployment/kubernetes-start.html.

2. Deploy the Sock Shop in ArgoCD. To deploy the app, you will need to do the following:

    I.   Create a new ArgoCD app.

    II. Point to the repo where the app exists.

    III. Point to the destination server, which is the server or service that you're running Kubernetes on.

    IV. Specify the destination namespace:

    ```
    argocd app create socks --repo https://github.com/
    microservices-demo/microservices-demo.git --path deploy/
    kubernetes --dest-server https://kubernetes.default.svc
    --dest-namespace sock-shop
    ```

3. Now that the app is deployed, you can check the status of the app:

    ```
    argocd app get socks
    ```

```
GROUP   KIND        NAMESPACE   NAME         STATUS    HEALTH       HOOK   MESSAGE
        Namespace   sock-shop   sock-shop    Running   Synced              namespace/sock-shop configured. Warning: resource name
aces/sock-shop is missing the kubectl.kubernetes.io/last-applied-configuration annotation which is required by  apply.  apply s
ould only be used on resources created declaratively by either  create --save-config or  apply. The missing annotation will be
atched automatically.
        Service     sock-shop   carts-db     Synced    Healthy             service/carts-db created
        Service     sock-shop   carts        Synced    Healthy             service/carts created
        Service     sock-shop   user-db      Synced    Healthy             service/user-db created
        Service     sock-shop   rabbitmq     Synced    Healthy             service/rabbitmq created
        Service     sock-shop   shipping     Synced    Healthy             service/shipping created
        Service     sock-shop   payment      Synced    Healthy             service/payment created
        Service     sock-shop   queue-master Synced    Healthy             service/queue-master created
        Service     sock-shop   catalogue    Synced    Healthy             service/catalogue created
        Service     sock-shop   session-db   Synced    Healthy             service/session-db created
        Service     sock-shop   orders       Synced    Healthy             service/orders created
        Service     sock-shop   catalogue-db Synced    Healthy             service/catalogue-db created
        Service     sock-shop   orders-db    Synced    Healthy             service/orders-db created
        Service     sock-shop   front-end    Synced    Healthy             service/front-end created
        Service     sock-shop   user         Synced    Healthy             service/user created
apps    Deployment  sock-shop   carts-db     Synced    Progressing         deployment.apps/carts-db created
apps    Deployment  sock-shop   carts        Synced    Progressing         deployment.apps/carts created
apps    Deployment  sock-shop   user         Synced    Progressing         deployment.apps/user created
apps    Deployment  sock-shop   user-db      Synced    Progressing         deployment.apps/user-db created
apps    Deployment  sock-shop   catalogue-db Synced    Progressing         deployment.apps/catalogue-db created
apps    Deployment  sock-shop   rabbitmq     Synced    Progressing         deployment.apps/rabbitmq created
apps    Deployment  sock-shop   orders       Synced    Progressing         deployment.apps/orders created
apps    Deployment  sock-shop   front-end    Synced    Progressing         deployment.apps/front-end created
apps    Deployment  sock-shop   payment      Synced    Progressing         deployment.apps/payment created
apps    Deployment  sock-shop   session-db   Synced    Progressing         deployment.apps/session-db created
apps    Deployment  sock-shop   queue-master Synced    Progressing         deployment.apps/queue-master created
apps    Deployment  sock-shop   shipping     Synced    Progressing         deployment.apps/shipping created
apps    Deployment  sock-shop   orders-db    Synced    Progressing         deployment.apps/orders-db created
apps    Deployment  sock-shop   catalogue    Synced    Progressing         deployment.apps/catalogue created
        Namespace               sock-shop    Synced
```

Figure 6.12 – Sock Shop resources

You can now check that the app was deployed in the ArgoCD UI.

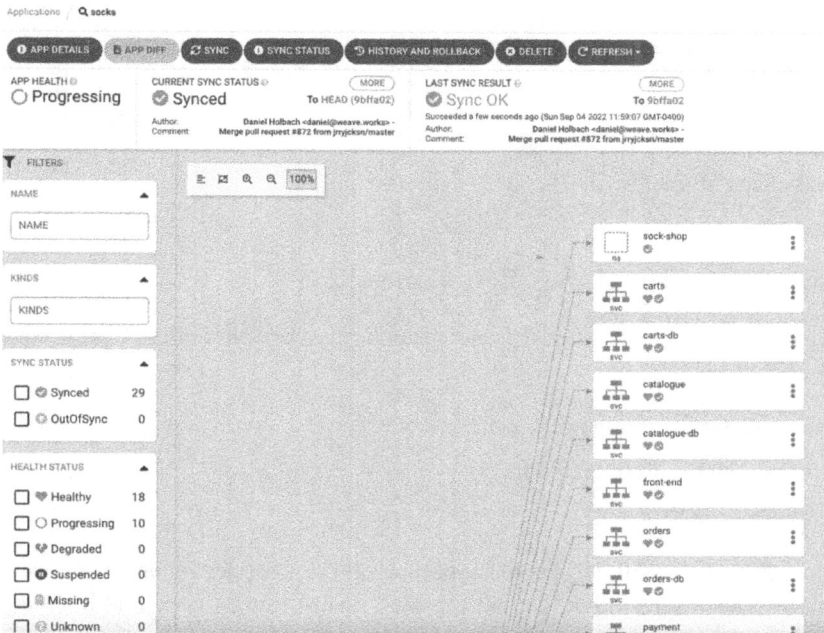

Figure 6.13 – Sock Shop app connection

You'll see the health of the app, whether it's synced, and whether the status of the application is as expected.

Production use cases for CI/CD and GitOps

Two ways to think about CI/CD and GitOps in production are as follows:

- CI/CD should be used to deploy the cluster
- GitOps should be used to manage the Kubernetes resources inside the cluster

In other words, CI/CD deploys the infrastructure and clusters and GitOps deploys and manages the apps. Use the best tool for the job, which is the infrastructure deployment type of workflow.

Regardless of which GitOps and CI/CD solution you use, you always want to keep in mind that your goal is to automate and create repeatable workflows that work for you and your team. Regardless of what *hot* tool or platform is out right now, you want to use what's best for your team, not whatever is the *new thing*.

In the next section, you'll dive into multiple methods of troubleshooting containerized apps running in your Kubernetes cluster.

Troubleshooting application deployments

Troubleshooting environments and applications typically always looks the same and follows a typical order:

- When was the last deployment?
- What has changed?
- Look at the logs
- Who can access the app and who cannot, if anyone at all?

With Kubernetes, it's pretty similar when it comes to application troubleshooting. The usual workflow is as follows:

1. Check the app itself running in the container.
2. Check the overall health of the Pod(s).
3. Check the Service or route.

With these three steps, you can usually get to the bottom of what's happening because, in reality, there can't be any other problems. It's either that the app itself isn't working, the Pod itself has an issue, or the service or route isn't working as expected.

Although there could only be three potential problems at a high level, when you dive deeper into those problems, there could be various ways to troubleshoot the current issue you're facing, which you'll learn about in this section.

As with all troubleshooting techniques, you should think about it in the following order:

- What's the problem?
- What's changed?
- What could be the problem in the problem? As in, a Pod may be down, but it might not be because of the app. It could be because of a problem with the replication controller.

Troubleshooting Pods

The two commands that'll help you debug Pods are as follows:

- `kubectl describe`
- `kubectl logs`

Take the following Kubernetes Manifest and deploy it. Notice how, for the container tag, it's spelled as `lates`. That's on purpose, as you want the container to fail:

```
apiVersion: apps/v1
kind: Deployment
metadata:
  name: nginx-deployment
spec:
  selector:
    matchLabels:
      app: nginxdeployment
  replicas: 2
  template:
    metadata:
      labels:
        app: nginxdeployment
    spec:
      containers:
      - name: nginxdeployment
        image: nginx:lates
```

```
        ports:
        - containerPort: 80
```

Retrieve the name of the Pod with the following command:

```
kubectl get pods
```

You'll see an output similar to the following:

```
NAME                                   READY   STATUS         RESTARTS   AGE
nginx-deployment-78bb975ccb-ll6nt      0/1     ErrImagePull   0          12s
nginx-deployment-78bb975ccb-wl898      0/1     ErrImagePull   0          12s
```

Figure 6.14 – Error container image pull

Notice how, right off the bat, you can start the troubleshooting process. The status states that there was an error pulling the image. Now you know that there's an issue with the image, let's dive a bit deeper.

Run the following command:

```
kubectl describe pods pod_name
```

You'll see an output similar to the following screenshot:

```
Conditions:
  Type              Status
  Initialized       True
  Ready             False
  ContainersReady   False
  PodScheduled      True
Volumes:
  kube-api-access-9f4gz:
    Type:                     Projected (a volume that contains injected data from multiple sources)
    TokenExpirationSeconds:   3607
    ConfigMapName:            kube-root-ca.crt
    ConfigMapOptional:        <nil>
    DownwardAPI:              true
QoS Class:                    BestEffort
Node-Selectors:               <none>
Tolerations:                  node.kubernetes.io/not-ready:NoExecute op=Exists for 300s
                              node.kubernetes.io/unreachable:NoExecute op=Exists for 300s
Events:
  Type     Reason     Age                   From               Message
  ----     ------     ----                  ----               -------
  Normal   Scheduled  3m1s                  default-scheduler  Successfully assigned default/nginx-deployment-78bb975ccb-ll6nt to minikube
  Normal   Pulling    83s (x4 over 3m)      kubelet            Pulling image "nginx:lates"
  Warning  Failed     83s (x4 over 3m)      kubelet            Failed to pull image "nginx:lates": rpc error: code = Unknown desc = Error
  response from daemon: manifest for nginx:lates not found: manifest unknown: manifest unknown
  Warning  Failed     83s (x4 over 3m)      kubelet            Error: ErrImagePull
  Warning  Failed     71s (x6 over 2m59s)   kubelet            Error: ImagePullBackOff
  Normal   BackOff    59s (x7 over 2m59s)   kubelet            Back-off pulling image "nginx:lates"
```

Figure 6.15 – Pod description

The great thing about the `describe` command is that it gives you a log output underneath the `Events` section. You can now see that the issue is that it couldn't pull the container image based on the name and tag that you gave.

The last step would be to run the `logs` command to see whether there's any other data you can use:

```
kubectl logs pod_name
```

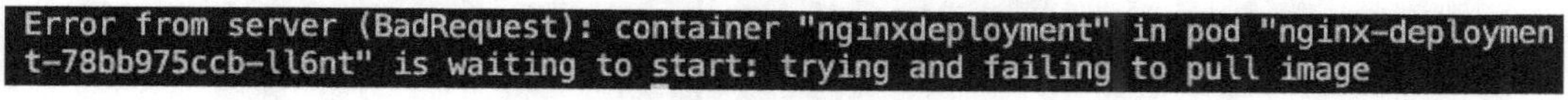

```
Error from server (BadRequest): container "nginxdeployment" in pod "nginx-deploymen
t-78bb975ccb-ll6nt" is waiting to start: trying and failing to pull image
```

Figure 6.16 – Pod logs

You can see from the screenshot here that there isn't much more to go off of other than what was given in the `describe` command, so the troubleshooting has been successfully completed.

Troubleshooting Services

When troubleshooting Services, the first thing that you always want to confirm is whether the Service exists. If you don't have a Service running in a Kubernetes cluster, you can use this example Manifest:

```yaml
apiVersion: apps/v1
kind: Deployment
metadata:
  name: nginx-deployment
spec:
  selector:
    matchLabels:
      app: nginxdeployment
  replicas: 2
  template:
    metadata:
      labels:
        app: nginxdeployment
    spec:
      containers:
      - name: nginxdeployment
        image: nginx:latest
        ports:
        - containerPort: 80
---
apiVersion: v1
kind: Service
metadata:
```

```
   name: nginxservice
spec:
  selector:
    app: nginxdeployment
  ports:
    - protocol: TCP
      port: 80
  type: LoadBalancer
```

Because Pod networks are separate from a host network, you'll need a Pod to exec or SSH into so you can do the troubleshooting. The following is a command that you can use to configure a Pod for troubleshooting purposes based on a busybox container image, which is a popular image used for troubleshooting purposes:

```
kubectl run -it --rm --restart=Never busybox --image=gcr.io/
google-containers/busybox sh
```

First, see whether the service is running. You'll do this outside of the busybox container image:

```
kubectl get service
```

You should get the following output.

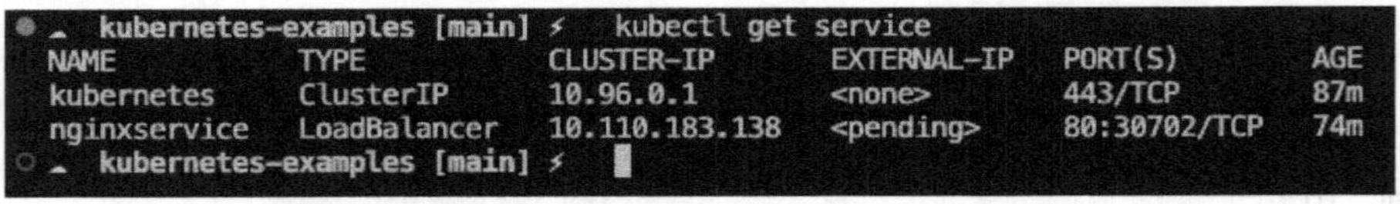

Figure 6.17 – Service configuration

If the service is running, confirm that you can reach the service via DNS:

```
nslookup service_name
```

You'll see an output similar to the following configuration:

Figure 6.18 – nslookup of Pod

If the standard `nslookup` command doesn't work, or if you want another type of confirmation, try an FQDN:

```
nslookup service_name.namespace_name.svc.cluster.local
```

You'll see an output similar to the following screenshot.

```
/ # nslookup nginxservice.default.svc.cluster.local
Server:     10.96.0.10
Address 1:  10.96.0.10 kube-dns.kube-system.svc.cluster.local

Name:       nginxservice.default.svc.cluster.local
Address 1:  10.110.183.138 nginxservice.default.svc.cluster.local
/ #
```

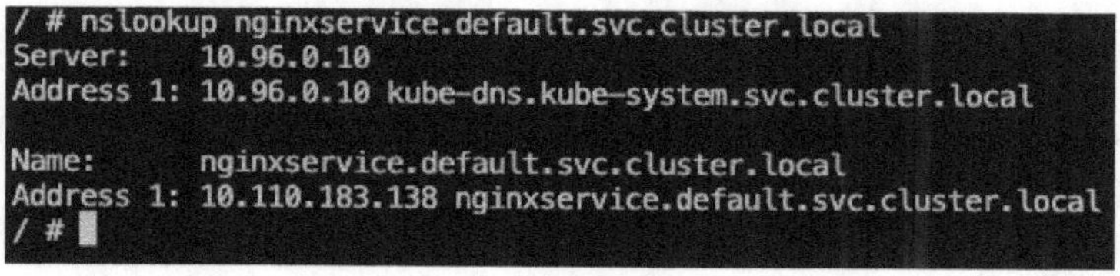

Figure 6.19 – FQDN service lookup

Check to confirm that the service is defined correctly:

```
kubectl get service name_of_service -o json
```

You'll see an output similar to the following screenshot.

```
kubernetes-examples [main]  ⚡  kubectl get service nginxservice -o json
{
    "apiVersion": "v1",
    "kind": "Service",
    "metadata": {
        "annotations": {
            "kubectl.kubernetes.io/last-applied-configuration": "{\"apiVersion\":\"v1\",\"kind\":\"Service\",\"metadata\":{\"annotations\":
},\"name\":\"nginxservice\",\"namespace\":\"default\"},\"spec\":{\"ports\":[{\"port\":80,\"protocol\":\"TCP\"}],\"selector\":{\"app\":\"ngi
nxdeployment\"},\"type\":\"LoadBalancer\"}}\n"
        },
        "creationTimestamp": "2022-09-09T12:43:26Z",
        "name": "nginxservice",
        "namespace": "default",
        "resourceVersion": "1068",
        "uid": "096e2852-d486-4d2d-a34a-8952f38acbc5"
    },
    "spec": {
        "allocateLoadBalancerNodePorts": true,
        "clusterIP": "10.110.183.138",
        "clusterIPs": [
            "10.110.183.138"
        ],
        "externalTrafficPolicy": "Cluster",
        "internalTrafficPolicy": "Cluster",
        "ipFamilies": [
            "IPv4"
        ],
        "ipFamilyPolicy": "SingleStack",
        "ports": [
            {
                "nodePort": 30702,
                "port": 80,
```

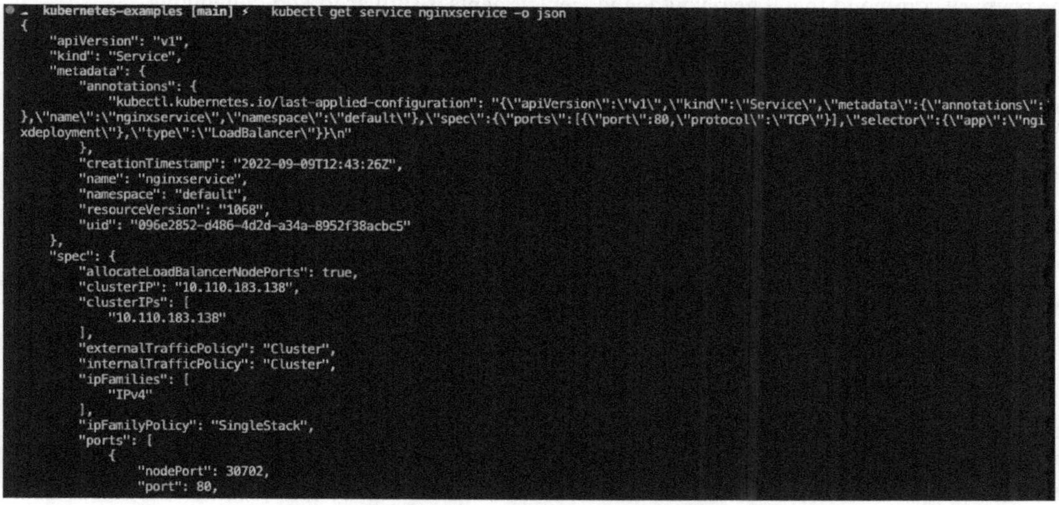

Figure 6.20 – JSON output of service

Check that the service has endpoints – as in, confirm that there are Pods that the service is pointing to:

```
kubectl get pods -l app=name_of_deployment
```

You'll see an output similar to the following screenshot:

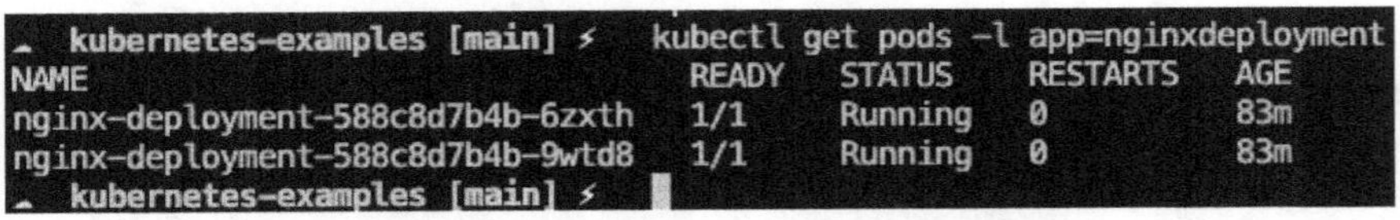

```
 kubernetes-examples [main]     kubectl get pods -l app=nginxdeployment
NAME                               READY   STATUS    RESTARTS   AGE
nginx-deployment-588c8d7b4b-6zxth   1/1     Running   0          83m
nginx-deployment-588c8d7b4b-9wtd8   1/1     Running   0          83m
 kubernetes-examples [main]
```

Figure 6.21 – Retrieving Pods based on label

Finally, which should already be known, but just in case, check to confirm that the Pods that the service is pointing to are working:

```
kubectl get pods
```

The final piece, which you'll learn about in the next section, is implementing a service mesh for troubleshooting. A service mesh has several jobs, and one of the jobs is making it easier to troubleshoot latency issues between Services, along with ensuring that Services are working as expected.

Troubleshooting Deployments

The primary command that'll help you debug Deployments is similar to Pod debugging:

```
kubectl describe deployment
```

Unless there's something wrong with the Deployment controller itself, there usually isn't a problem with the actual Deployment. It's typically a problem with the Pods running inside of the Deployment. However, you still may want to check the Deployment itself.

To do that, you would run the following:

```
kubectl describe deployment deployment_name
```

You should get an output similar to the following screenshot:

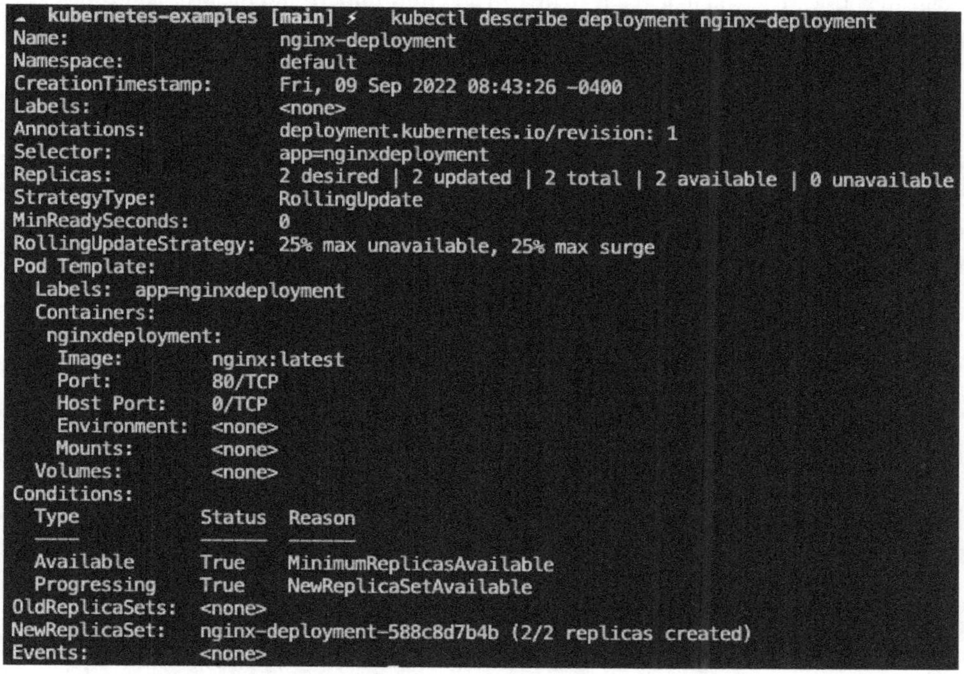

```
  kubernetes-examples [main]    kubectl describe deployment nginx-deployment
Name:                    nginx-deployment
Namespace:               default
CreationTimestamp:       Fri, 09 Sep 2022 08:43:26 -0400
Labels:                  <none>
Annotations:             deployment.kubernetes.io/revision: 1
Selector:                app=nginxdeployment
Replicas:                2 desired | 2 updated | 2 total | 2 available | 0 unavailable
StrategyType:            RollingUpdate
MinReadySeconds:         0
RollingUpdateStrategy:   25% max unavailable, 25% max surge
Pod Template:
  Labels:   app=nginxdeployment
  Containers:
   nginxdeployment:
    Image:        nginx:latest
    Port:         80/TCP
    Host Port:    0/TCP
    Environment:  <none>
    Mounts:       <none>
  Volumes:        <none>
Conditions:
  Type           Status  Reason
  ----           ------  ------
  Available      True    MinimumReplicasAvailable
  Progressing    True    NewReplicaSetAvailable
OldReplicaSets:  <none>
NewReplicaSet:   nginx-deployment-588c8d7b4b (2/2 replicas created)
Events:          <none>
```

Figure 6.22 – Describing the nginx Deployment

The goal of the describe command isn't to tell you about logs or events or what's happening – it's to help you fully understand what's deployed and how it's deployed. That way, you can backtrack and see whether what's deployed is actually supposed to be there.

In the next section, we'll wrap up this chapter by talking about what a service mesh is, what an Ingress is, and how to think about implementing them.

Service meshes and Ingresses

Almost every containerized application needs to be routed in one way or another – whether it's so outside users can use the application, applications can talk to each other, or one application needs to connect to another. Routes and Services are extremely important in Kubernetes, which is why service meshes and Ingresses play a huge part.

In many cases, you'll need better visuals into what's happening with services, how traffic is being routed, and what applications are routing to which load balancers and IP addresses. You'll also eventually want a way to encrypt traffic between services, which Kubernetes doesn't give you out of the box.

Service meshes and Ingresess are typically more advanced-level topics, but in this book and possibly at this stage in your career, you'll be ready to dive in and fully understand the pros and cons of using these two tools, plugins, and platforms.

Why Ingress?

At this point in your Kubernetes journey, it's almost certain that you've seen a Kubernetes Service. In fact, you've seen them throughout this book. A lot of the time, a Kubernetes Service comes with a frontend app that's attached to it, in which you need a way for users to interact with the Kubernetes Service. It's typically in front of a load balancer.

The problem with that is if you have a load balancer in front of your service, you have to do the following:

- Pay extra for the cloud load balancer if you're using a cloud Kubernetes service

- Set up a virtual load balancer if you're using an on-premises Kubernetes cluster

- Have a bunch of load balancers to manage

With an Ingress controller, you don't have to worry about that.

You can have several different Kubernetes Services and have an Ingress controller point to all of them, and each of the services can be reached by a different path.

Ingress controllers save time, money, management, and effort for engineers.

Using Ingresses

Now that you know about Ingress controllers, let's see how one can be configured using (at the time of writing) the most popular option, Nginx Ingress.

First things first – you'll need a Kubernetes Deployment and Service to deploy. If you don't already have them and would like to keep things simple, you can use the following Kubernetes Manifest, which is a sample Azure app. It doesn't have to be running in Azure to work:

```
apiVersion: apps/v1
kind: Deployment
metadata:
  name: aks-helloworld-one
spec:
  replicas: 1
  selector:
    matchLabels:
      app: aks-helloworld-one
  template:
```

```
      metadata:
        labels:
          app: aks-helloworld-one
      spec:
        containers:
        - name: aks-helloworld-one
          image: mcr.microsoft.com/azuredocs/aks-helloworld:v1
          ports:
          - containerPort: 80
          env:
          - name: TITLE
            value: "Welcome to Azure Kubernetes Service (AKS)"
---
apiVersion: v1
kind: Service
metadata:
  name: aks-helloworld-one
spec:
  type: ClusterIP
  ports:
  - port: 80
  selector:
    app: aks-helloworld-one
```

Once the app itself is deployed, you can deploy the Ingress controller. The Ingress controller is part of the native Kubernetes API set from the named group, so you don't have to worry about installing other CRDs or controllers:

```
apiVersion: networking.k8s.io/v1
kind: Ingress
metadata:
  name: hello-world-ingress-static
  annotations:
    nginx.ingress.kubernetes.io/ssl-redirect: "false"
spec:
  ingressClassName: nginx
  rules:
```

```
    - http:
        paths:
        - path: /
          pathType: Prefix
          backend:
            service:
              name: aks-helloworld-one
              port:
                number: 80
```

The last step is to forward to the port for the app's service so you can reach the app locally:

```
kubectl port-forward service/aks-helloworld-one :80
```

You should get an output similar to the following screenshot:

```
▴ aks [main] ⚡   kubectl port-forward service/aks-helloworld-one :80
Forwarding from 127.0.0.1:55319 -> 80
Forwarding from [::1]:55319 -> 80
```

Figure 6.23 – Port-forwarding service communication

You should be able to reach out to the app over localhost.

Welcome to Azure Kubernetes Service (AKS)

Figure 6.24 – AKS app

Now that you know what an Ingress is from a theoretical and practical perspective, let's move on to service meshes and look at how communication can occur more securely.

Why service meshes?

When you deploy containerized applications into a Kubernetes cluster, there are two primary ways that those applications communicate:

- Services
- Pod-to-Pod communication

Pod-to-Pod communication isn't recommended because Pods are ephemeral, which means they aren't permanent. They are designed to go down at any time and only if they were part of a StatefulSet would they keep any type of unique identifier. However, Pods still need to be able to communicate with each other. Backends need to talk to frontends, middleware needs to talk to backends and frontends, and so on.

The next communication method, which is the primary, is service-to-service. Service-to-service is the preferred method because a Service isn't ephemeral and only gets deleted if manually deleted. Pods can connect to Services with Selectors or Tags. If a Pod goes down, but the Selector in the Kubernetes Manifest that deployed the Pod doesn't change, the new Pod will be connected to the Service.

Here's the primary concern with everything described so far – all this traffic is unencrypted. Pod-to-Pod communication, or as some people like to call it, East-West traffic, is unencrypted. That means if for any reason a Pod is compromised or you have some segregation issues, there's nothing out of the box that you can do.

That's where a service mesh comes into play. A service mesh has the ability to do the following:

- Encrypt traffic between microservices
- Help with network latency troubleshooting
- Securely connect Kubernetes Services
- Perform observability for tracing and alerting

Using a service mesh

Now that you know about a service mesh, let's learn how to set one up. There are a ton of different service mesh platforms out there, all of which have their own method of being installed and configured.

Because it's a complicated topic in itself, there's no way to get through it all in this section. In fact, there are literal books for just service meshes. Let's learn how to set up an Istio service mesh.

First, download Istio:

```
curl -L https://istio.io/downloadIstio | sh
```

Next, export the path to the $PATH variable:

```
export PATH=$PWD/bin:$PATH
```

Output the $PATH variable int, bashrc:

```
echo "export PATH=$PATH:$HOME/istio-1.15.0/bin" >> ~/.bashrc
```

Install Istio on your Kubernetes cluster. Notice how Ingress is set to false. You can set it to true if you want to use the Istio Ingress. If you're using another Ingress controller, such as Nginx Ingress, you can leave it as false:

```
istioctl install --set values.gateways.istio-ingressgateway.
enabled=false
```

Istio is a great service mesh but doesn't have a UI out of the box. One of the most popular ways to look at your service mesh graphically is by using Kiali, which is a simple install:

```
kubectl apply -f https://raw.githubusercontent.com/istio/istio/
release-1.15/samples/addons/kiali.yaml
```

Set up port forwarding to Kiali so you can reach the UI locally:

```
kubectl port-forward -n istio-system service/kiali :20001
```

The last step is to take a Kubernetes Manifest, like the one you used in this chapter, and inject the sidecar (the service mesh container) into your Kubernetes Deployment:

```
istioctl kube-inject -f nginx.yaml | kubectl apply -f -
```

At this point, you now have the theoretical grounding and a bit of hands-on knowledge for how to move forward on your service mesh journey.

Summary

Overall, containerized application deployment, troubleshooting, and third-party tooling are going to be the core pieces of what your Kubernetes cluster looks like. Without proper troubleshooting, you won't have successful deployments. Without third-party tooling such as Ingress controllers, you won't be able to properly manage frontend apps. Out of the box, Kubernetes gives you a ton to use to make things work. However, there are more steps you need to take. For better or for worse, Kubernetes isn't one of those platforms where you can just deploy it and walk away. It takes management and engineering skills to ensure it's working as expected.

In the next chapter, you'll learn about how to monitor the workloads you've been deploying throughout this book.

Further reading

- *Learn Helm* by Andrew Block and Austin Dewey: `https://www.packtpub.com/product/learn-helm/9781839214295`

- *Mastering Service Mesh* by Anjali Khatri and Vikram Khatri: `https://www.packtpub.com/product/mastering-service-mesh/9781789615791?_ga=2.161313023.37784508.1672298745-664251363.1663254593`

Part 3: Final 15 Kubernetes Concepts – Security and Monitoring

Throughout this book so far, you've learned about key critical pieces of Kubernetes. First, you learned about various cluster methods. You learned about on-premises, how to deploy various managed Kubernetes services, and the different options that are available when deploying.

After that, you learned about different deployment methods, both from a *basic* perspective and an *advanced* perspective.

The only things that are missing, which are arguably the most important pieces to a proper Kubernetes environment, are monitoring, observability, and security. The third and final piece of the 50 concepts will go over what monitoring and observability look like in today's world.

In the last two chapters, you'll learn all about monitoring, observability, and security. You'll dive into the theory behind monitoring and observability, along with the hands-on portions of each so you can get a true understanding of how to implement different tools. Next, you'll dive into security, which is typically overlooked in general but especially overlooked in Kubernetes.

By the end of these, you'll have a solid theoretical overview of monitoring, observability, and security, along with real code examples for implementing the different tools and platforms available.

This part of the book comprises the following chapters:

- *Chapter 7, Kubernetes Monitoring and Observability*
- *Chapter 8, Security Reality Check*

7

Kubernetes Monitoring and Observability

Monitoring and observability for both Ops and Dev teams have always been crucial. Ops teams used to be focused on infrastructure health (virtual machines, bare-metal, networks, storage, and so on) and Devs used to be focused on application health. With Kubernetes, those lines are blurred. In a standard data center environment, it's easy to split who's conducting monitoring and observability in a very traditional sense. Kubernetes blends those lines because, for example, Pods are, in a sense, infrastructure pieces because they have to scale and are sort of *virtual machines* in the traditional sense. They are what holds the application. However, the application is running in a Pod, so if you're monitoring a Pod, you're automatically monitoring the containers that are running inside of the Pod.

Because these lines are blurred, both teams are doing both parts of the monitoring process. On a platform engineering or DevOps engineering team, those teams would monitor both application pieces and infrastructure pieces.

There's no longer a line that's used to divide which team monitors and creates observability practices around specific parts of Kubernetes. Instead, the goal is now to have a more unified front to ensure that the overall environment and applications are working as expected.

In this chapter, you're going to dive in from a theoretical and hands-on perspective to truly get an understanding of monitoring and observability in Kubernetes. The goal is for you to be able to take what you learn and what you've implemented in your lab from this chapter and truly start to use it in production. First, you'll learn what monitoring and observability actually are. Next, you'll learn what monitoring and observability mean for the infrastructure layer, which is the virtual machines running the Kubernetes environment, and the specifics around Control Plane and worker node monitoring. After that, you'll dive into monitoring and observability for specific Kubernetes resources such as Pods and Services. To finish up, you'll look at specific tools and platforms that are typically used in today's world for monitoring and observability.

Without monitoring, engineers wouldn't know what's happening inside a system or application. It's the job of a DevOps and platform engineer to have that information and make good use of that information by fixing whatever is broken.

In this chapter, we're going to cover the following main topics:

- How monitoring is different than observability
- Monitoring and observability tools for Kubernetes
- Observability practices
- Kubernetes resource monitoring

Technical requirements

This chapter isn't going to be a full-blown explanation of monitoring. Although there will be some brief explanations as a refresher/starting point, it's important that you have some experience in monitoring and observability. For example, maybe you've used the Kubernetes Dashboard before or you've looked at pre-populated monitors inside of Azure or AWS. It could even be experience monitoring from your local desktop.

You can find the GitHub repo here: `https://github.com/PacktPublishing/50-Kubernetes-Concepts-Every-DevOps-Engineer-Should-Know/tree/main/Ch7/prometheus/helm`

How is monitoring different than observability?

Two of the closest workflows and the two that are most often interchanged from a verbiage and explanation perspective are monitoring and observability. Although this chapter isn't dedicated to observability, to truly understand the differences between monitoring and observability, you must understand both and ultimately see how they work. After the explanations in this section, you'll see that there are key differences between observability and monitoring, along with differences in how they should be used, when they should be used, and the best practices for them.

What you might experience in organizations, depending on how mature their engineering teams are, is that monitoring and observability get thrown into one category. They are both either looked at the same way, or engineering teams think they're doing observability when really all they're doing is monitoring. One of the goals of this chapter is to give you the ability to differentiate between the two because there can be some blurred lines depending on what platforms and tools you're using. For example, let's take two of the most popular platforms – Datadog and New Relic. Both of these platforms are looked at as monitoring platforms and observability platforms. They can both do monitoring and observability, and they do them well. This is not always the case though. A platform such as Prometheus is just for observability and collecting metrics, but you can pair it with a monitoring platform/tool such as Grafana to give you a visual of what's happening inside of an environment.

Monitoring and observability are both lengthy topics, especially in Kubernetes. The way that monitoring and observability are thought of in Kubernetes is similar to other platforms and systems, but vastly different.

In the next section, you're going to look at what monitoring and observability are and how to know which you should use. We'll also explore a few monitoring versus observability examples.

What's monitoring?

Have you ever opened up **Task Manager** in Windows, gone to the performance settings, and looked at the memory and/or CPU usage? What about **Activity Monitor** on macOS to see what applications and programs were using memory and CPU? If you've done either of these things, which it is safe to assume that most engineers have done at one point or another, you've officially monitored a system! Now, you may be thinking to yourself that checking out the memory and CPU on a desktop or laptop is drastically different, but it's actually not. Regardless of whether it's a desktop or an entire server rack, RAM is RAM, CPU is CPU, and storage is storage. It doesn't change across systems. The only thing that changes is the amount of CPU, memory, and storage.

So, what is monitoring?

Monitoring is the ability to view system resources, performance, and usage in real time. You can monitor anything in a Kubernetes cluster including the following:

- Worker nodes
- Control Planes
- Pods
- Deployments
- ConfigMaps

As well as these, you can also monitor literally any other Kubernetes resource that's running in your cluster. From the application level to the infrastructure level to the networking level, it can all be monitored.

With monitoring can come the creation of alerts. I remember when I first got into tech and got my first internship, the coolest thing to me was walking into a **network operations center** (**NOC**) and seeing all the big screens with all the monitors on them. It was like we were protecting nuclear launch codes. It was amazing to see that every single system could be watched so engineers could understand what was happening underneath the hood.

In today's world, engineers are still using things such as big monitors in a NOC, but with working from home and the remote world being the new norm, engineers are also logging in to monitoring platforms to view how systems are working. Engineers can log in to tools such as Datadog, CloudWatch, or Azure Monitor and see everything that's happening with every service.

Let's take a look at the screenshot in *Figure 7.1* from Azure. As you can see, there are a ton of monitoring options available.

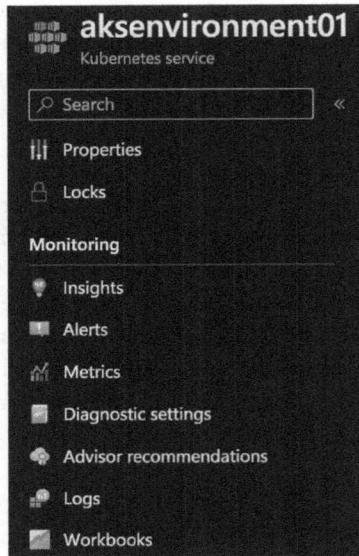

Figure 7.1 – The AKS monitoring options

The monitoring options that you see in the **Monitoring** section also contain some observability practices (such as **Metrics**), which goes back to a point made earlier in the chapter – there's some confusion when splitting up monitoring and observability practices.

From a monitoring perspective, what you should care about are the actual monitors.

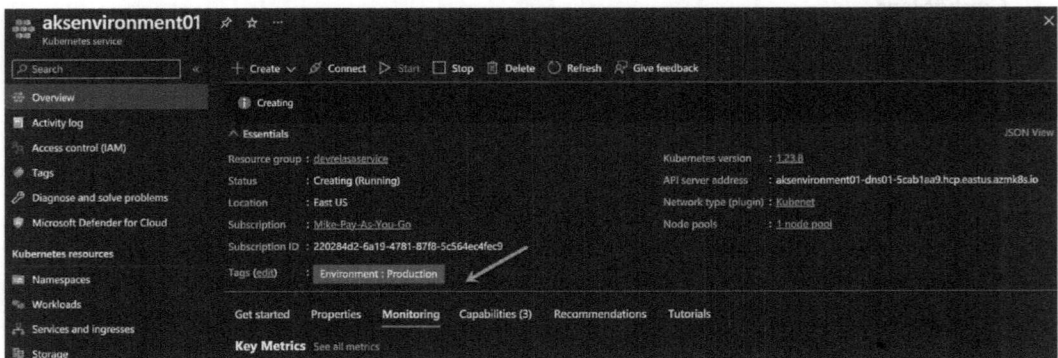

Figure 7.2 – AKS Monitoring

The monitoring information that you can pull from AKS, or nearly any other Azure service, gives you the ability to see what's happening right now or what's been happening for an extended period of time. This gives you the ability to understand how a system is performing but from an ad hoc perspective.

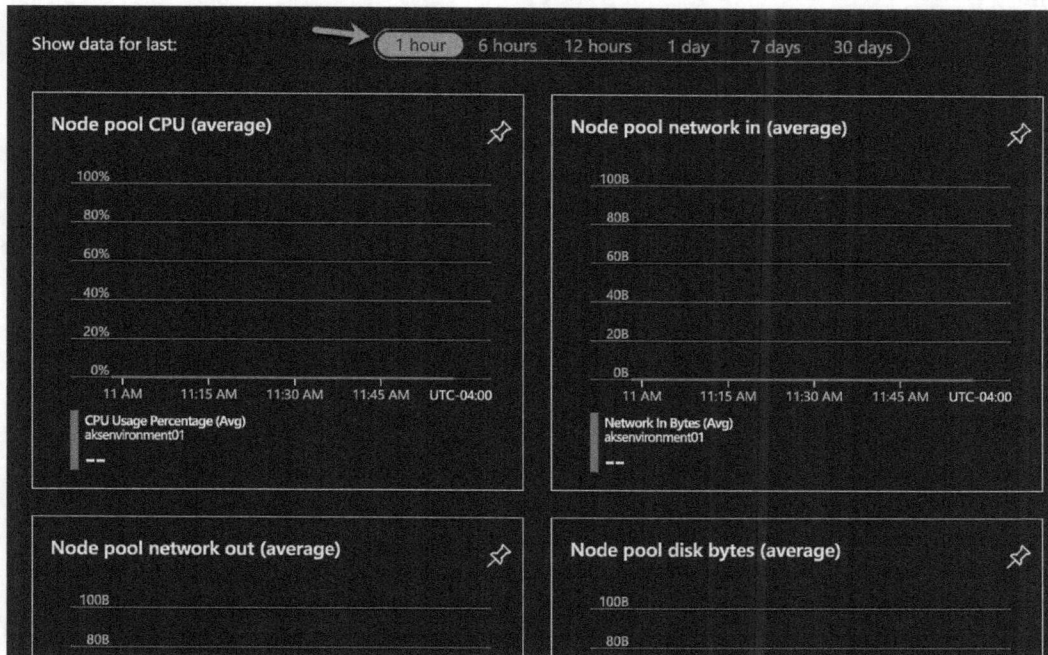

Figure 7.3 – The hardware metrics

The idea of this type of monitoring is to see and understand how cluster resources such as CPU, memory, storage, and bandwidth (inbound and outbound) are performing to ensure that you can make decisions about how a cluster should be managed.

You can also monitor applications that are running to see the uptime, how many resources they're consuming, and the overall performance of the apps.

Monitoring specifics on a Kubernetes cluster

The components on a Control Plane that you should monitor are the API server, etcd (the cluster store), controllers, and schedulers. The components on a worker node that you should monitor are Kubelet, container runtime, kube-proxy, and DNS. There's also the need to monitor Pods, but you'll be learning more about that at the end of this chapter.

In any circumstance, whether it's components on the Control Plane or components on the worker node, you should ensure that the Metrics Server is running. You can technically retrieve metrics via the `/metrics/resource` endpoint (example: `/metrics/pods`), but that would mean you have to query each resource. The Metrics Server goes to each resource, fetches the metrics, and exposes them instead of you having to retrieve them one by one. You can find the Metrics Server, which you can use across any Kubernetes cluster, here: `https://github.com/kubernetes-sigs/metrics-server/releases/latest/download/components.yaml`.

The Metrics Server endpoint comes from the Kubernetes **Specific Interest Group** (**SIG**) and can be deployed anywhere. Whether it's a Kubernetes cluster running in AWS or a Kubeadm cluster running on virtual machines on your Windows 10 laptop, it doesn't matter where the cluster exists.

What's the downside to monitoring?

The downside of monitoring, although it's powerful, is that there's not much that you can do with the data unless it's happening in real time. Sure, you can get alerts if there's an issue with a resource, but this means that an engineer would have to be on-call to fix the issue. They have to stop what they're doing to put out a fire. With the way that the tech world is going, this is not a sustainable model anymore.

Along with that, engineers want to spend more time creating value-driven work. They don't want to wake up at 2:00 A.M. due to getting an alert or stop coding a new feature because of an alert. Instead, they want a way to create automated and repeatable processes for an alert. For example, if an alert goes off, engineers want a way to create an automated process that can fix the problem if it happens. Then, they don't have to stop what they're doing to go put out a fire and can continue creating value-driven work.

This is where observability comes into play.

What's observability?

Because monitoring and observability are sometimes used interchangeably when explaining them, it's important to understand their differences. This way, as you dive deeper into monitoring, it's easier to understand the distinctions.

Observability is mostly what you'll see in Kubernetes and almost every other cloud-native system. However, monitoring and observability are starting to blend together in terms of what they mean. For example, in *Figure 7.1*, you saw the **Monitoring** section. Under the **Monitoring** section, there was a subsection for **Metrics**. The thing is, metrics technically fall under observability.

The reason why monitoring and observability are getting mashed together, or in other words, the reason why observability is becoming more popular, is that with observability, you can actually make decisions and automate workloads based on the data that you receive.

The key data points for observability practices are logs, metrics, and traces.

Again, we don't want to go too deep in this section because observability has an entire chapter to itself. Just remember three key things:

- Observability gives you the ability to perform an actual action with the data you're receiving. That action could be to automatically fix a resource that's causing problems.

- It's becoming increasingly popular over traditional monitoring.

- Observability has three key aspects: logs, metrics, and tracing.

A quick note on metrics

Metrics for most Kubernetes resources are exposed. They're exposed via the `/metrics/resource` endpoint. For example, `/metrics/pods` would be for the Pods Kubernetes resource.

To make things a bit easier, the Metrics Server, which isn't installed on Kubernetes out of the box (depending on the cloud provider, but out of the box means a raw Kubernetes cluster installation), can scrape and consolidate all of the metric endpoints for the Kubernetes resources. This way, you don't have to attempt to consume each metric via the resource one by one.

To kick things up a notch, there's the kube-state-metrics tool, which you can install on a Kubernetes server; its job is to focus on the health of the Kubernetes resources/objects on your cluster. For example, if the Pods are actually available and ready is what kube-state-metrics will look at and confirm.

If you're wondering what the difference is between the Metrics Server and kube-state-metrics, the Metrics Server shows cluster resource usage such as CPU and memory. On the other hand, kube-state-metrics is concerned with the health of the Kubernetes resource.

Monitoring versus observability examples

When thinking about how to implement monitoring, observability, or both, it's best to think about the implementation details from a scenario perspective.

Let's take two scenarios – one for a containerized application from a monitoring perspective and then taking the same containerized application, but looking at it from an observability perspective.

The following examples won't be a complete step-by-step guide. The code works, but it won't be explained in terms of how exactly to deploy and run it. Feel free to go through it on your own system, but the aim in this chapter is to show examples of the workflow rather than a complete step-by-step tutorial.

Monitoring use case

The first scenario can be thought about as, for example, a frontend application. It could be an Nginx web app, which is simple and hosts a website. It could be something as simple as the following Nginx configuration:

```
apiVersion: apps/v1
kind: Deployment
metadata:
  name: nginx-deployment
spec:
  selector:
    matchLabels:
      app: nginxdeployment
  replicas: 2
  template:
    metadata:
      labels:
        app: nginxdeployment
    spec:
      containers:
      - name: nginxdeployment
        image: nginx:latest
        ports:
        - containerPort: 80
```

With the preceding Kubernetes manifest, you can picture an application that's running with two replicas on a Kubernetes cluster. To retrieve the memory and CPU information of the Pod, you can run the `kubectl top` command:

kubectl top pod pod_name

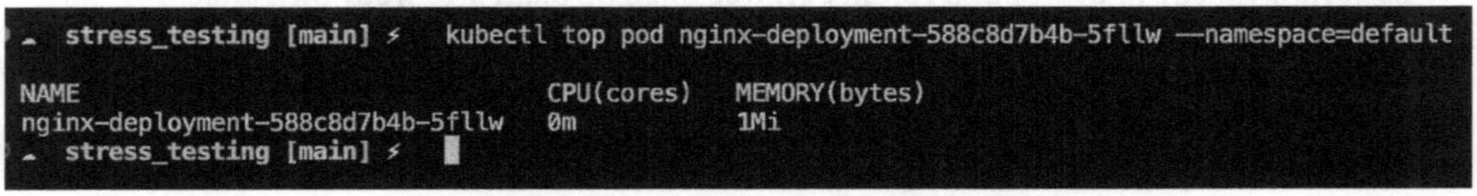

```
  stress_testing [main]  ⚡    kubectl top pod nginx-deployment-588c8d7b4b-5fllw —namespace=default
NAME                                CPU(cores)   MEMORY(bytes)
nginx-deployment-588c8d7b4b-5fllw   0m           1Mi
  stress_testing [main]  ⚡    █
```

Figure 7.4 – The top command

An error can sometimes occur if the Metrics API isn't enabled, as it's disabled by default. If you'd like to enable it, check the documentation for where you're running the Kubernetes cluster. As an example, here's how you'd enable the Metrics API on `minikube`:

```
minikube addons enable metrics-server
```

To stress-test the workload, you can use a stress/performance testing tool such as `k6`. The following is an example configuration that you can use:

```
import http from 'k6/http';
import { sleep } from 'k6';


export default function () {
  http.get('https://test.k6.io');
  sleep(1);
}
```

You can then save the preceding configuration and use it as a stress test with the following command, which specifies 100 virtual users and runs for 30 seconds:

```
k6 run --vus 100 --duration 30s test.js
```

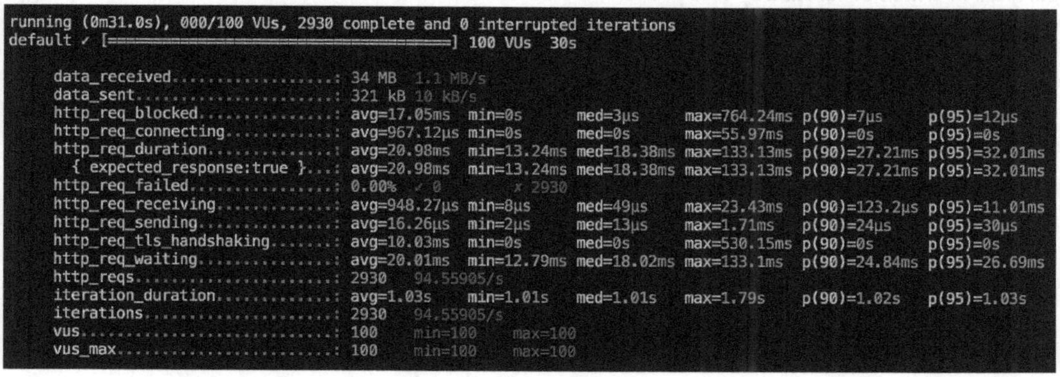

```
running (0m31.0s), 000/100 VUs, 2930 complete and 0 interrupted iterations
default ✓ [======================================] 100 VUs  30s

     data_received..................: 34 MB   1.1 MB/s
     data_sent......................: 321 kB  10 kB/s
     http_req_blocked...............: avg=17.05ms  min=0s      med=3µs    max=764.24ms p(90)=7µs     p(95)=12µs
     http_req_connecting............: avg=967.12µs min=0s      med=0s     max=55.97ms  p(90)=0s      p(95)=0s
     http_req_duration..............: avg=20.98ms  min=13.24ms med=18.38ms max=133.13ms p(90)=27.21ms p(95)=32.01ms
       { expected_response:true }...: avg=20.98ms  min=13.24ms med=18.38ms max=133.13ms p(90)=27.21ms p(95)=32.01ms
     http_req_failed................: 0.00%  ✓ 0         ✗ 2930
     http_req_receiving.............: avg=948.27µs min=8µs     med=49µs   max=23.43ms  p(90)=123.2µs p(95)=11.01ms
     http_req_sending...............: avg=16.26µs  min=2µs     med=13µs   max=1.71ms   p(90)=24µs    p(95)=30µs
     http_req_tls_handshaking.......: avg=10.03ms  min=0s      med=0s     max=530.15ms p(90)=0s      p(95)=0s
     http_req_waiting...............: avg=20.01ms  min=12.79ms med=18.02ms max=133.1ms  p(90)=24.84ms p(95)=26.69ms
     http_reqs......................: 2930   94.55905/s
     iteration_duration.............: avg=1.03s    min=1.01s   med=1.01s  max=1.79s    p(90)=1.02s   p(95)=1.03s
     iterations.....................: 2930   94.55905/s
     vus............................: 100    min=100     max=100
     vus_max........................: 100    min=100     max=100
```

Figure 7.5 – The benchmark test

Running the `kubectl top` command again, you can see that the memory increased:

```
 ▲  kubernetes-examples [main] ⚡   kubectl top pod nginx-deployment-588c8d7b4b-5fllw --namespace=default

NAME                                   CPU(cores)   MEMORY(bytes)  ←
nginx-deployment-588c8d7b4b-5fllw      0m           3Mi
```

Figure 7.6 – The kubectl top command for a Pod

After logging in to a piece of monitoring software, such as the Kubernetes Dashboard (which you'll learn about in the upcoming section), you will be able to see the CPU and memory utilization for both Pods.

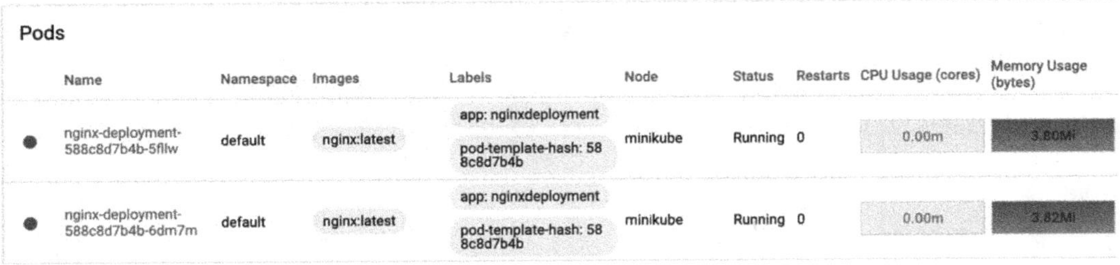

Pods									
	Name	Namespace	Images	Labels	Node	Status	Restarts	CPU Usage (cores)	Memory Usage (bytes)
●	nginx-deployment-588c8d7b4b-5fllw	default	nginx:latest	app: nginxdeployment pod-template-hash: 58 8c8d7b4b	minikube	Running	0	0.00m	3.80Mi
●	nginx-deployment-588c8d7b4b-6dm7m	default	nginx:latest	app: nginxdeployment pod-template-hash: 58 8c8d7b4b	minikube	Running	0	0.00m	3.82Mi

Figure 7.7 – The Pods running

This information gives you the ability to monitor what happens when more and more users access your application, which is very common for a frontend application.

Observability use case

The second scenario is going to be around checking out the Nginx Pods and Services that can be created from the Nginx configuration in the previous section. Ultimately, you'll be able to see how you can capture and view metrics data in an observability tool. Although *Figure 7.8* shows Prometheus, regardless of which observability tool you use, you're still going to see the same data because it's being retrieved via the Kubernetes Metrics API.

When the Metrics Server is enabled on a Kubernetes cluster, it exposes several resource metric endpoints. One of the resource metric endpoints is Pods. You can confirm that your Pod metrics are getting ingested into Prometheus based on **Service Discovery**.

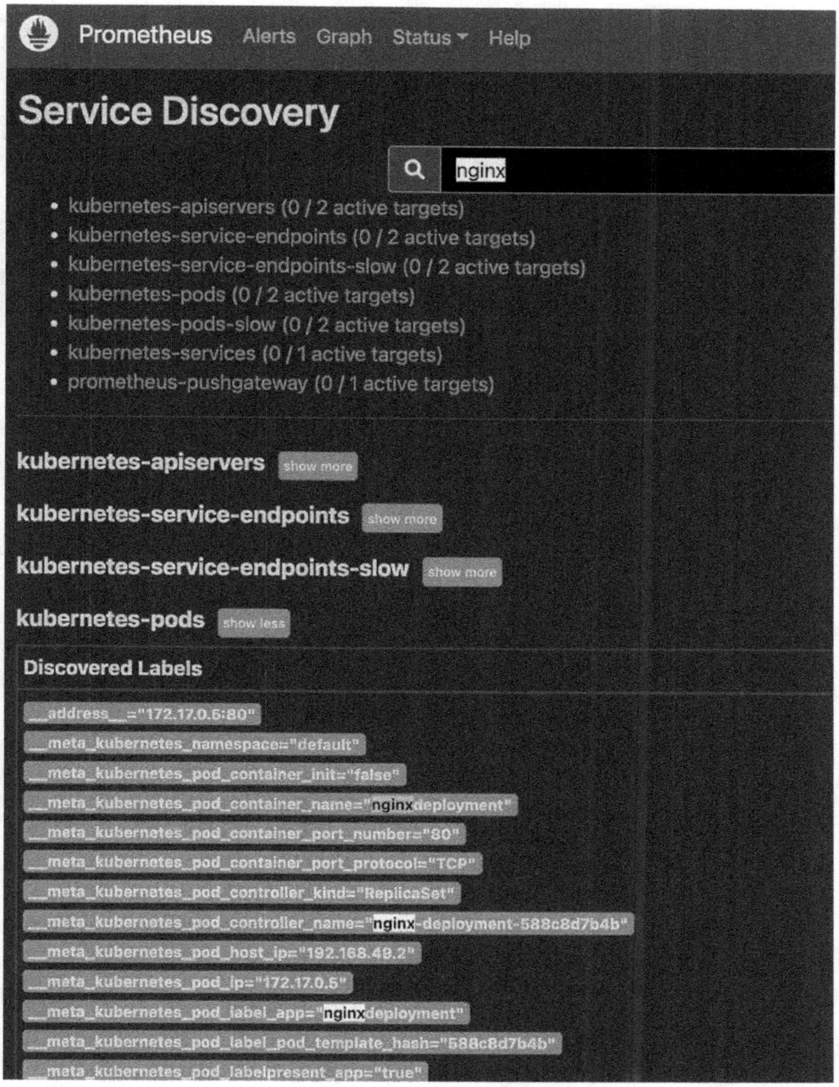

Figure 7.8 – A Pod discovery

You can then confirm how Pods are running based on different queries that Prometheus allows you to check with. For example, the following screenshot shows Kubernetes Service resource information, and you can see that the Nginx service is running.

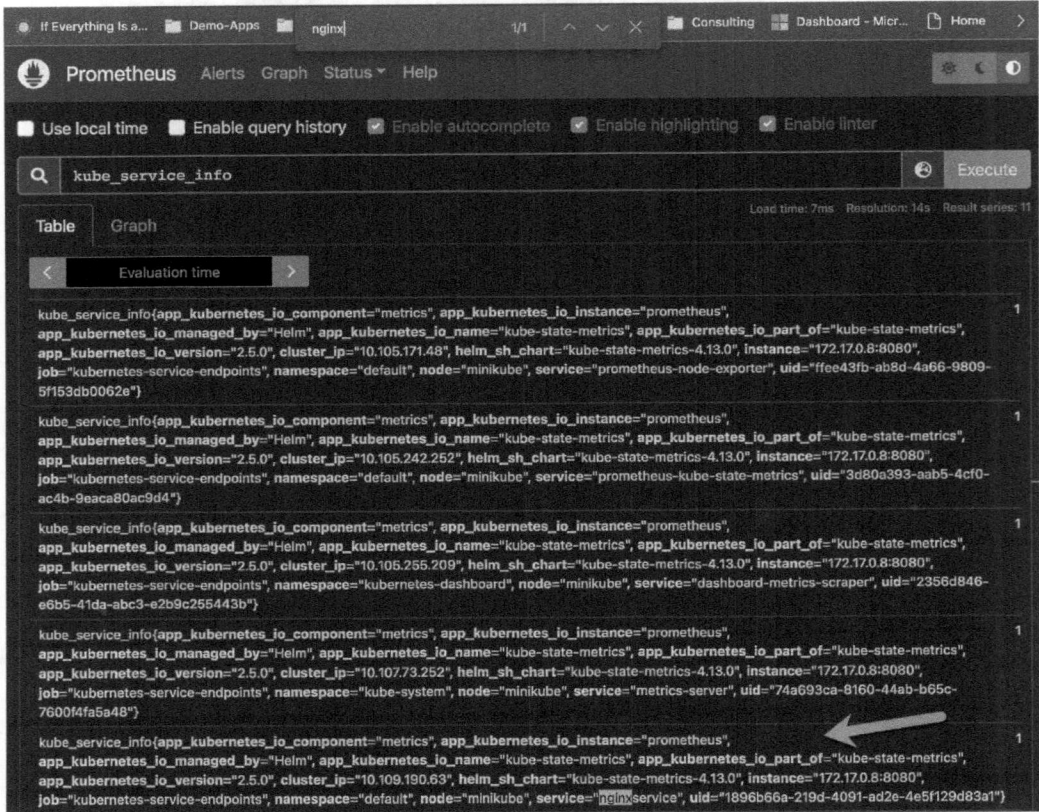

Figure 7.9 – Kubernetes Service metrics

You can also dive a little deeper and query based on certain hardware resources, such as memory and CPU. This way, you can understand how many resources (memory, CPU, and so on) are being taken up by each Pod.

For example, the following snippet is a query to see memory usage:

```
avg((avg (container_memory_working_set_bytes{pod="nginx-
deployment-588c8d7b4b-6dm7m"}) by (container_name , pod ))/ on
(container_name , pod)(avg (container_spec_memory_limit_bytes>0
) by (container_name, pod))*100)
```

Notice how a Pod name is specified; this will show you the observability metrics around memory for the specified Pod.

Monitoring and observability tools for Kubernetes

Typically, in any tech book, the theory/practical knowledge comes first, then the tooling. However, monitoring and observability are a bit different because you can't really talk about the specifics without mentioning or showing a certain tool/platform. Because of this, prior to jumping into the specifics around *how* to monitor and implement observability, you're going to learn about a few key tools.

The goal of this section is to help you first understand what the tools look like and then take the theory that you learn and utilize it in the tools. When you combine the knowledge and visuals (UI) of the tools with the understanding of what true monitoring and observability are, you can successfully implement them in your environment.

One of the interesting things about monitoring is that you can fully understand it from a theoretical perspective, but implementing it can be a challenge. For example, you can understand what the metrics endpoint in Kubernetes is, how it works, what metrics are exposed, and what resources you can monitor from those metrics. However, actually setting up a platform to *listen* to the metrics and configuring that listener is vastly different than reading about how metrics work.

Although this section won't cover all the tools and platforms used to monitor Kubernetes, this list is a great place to start as they are the most widely used in organizations. The good news is that even if you come across a monitoring tool that isn't covered in this section, monitoring is monitoring. That means once you understand monitoring and how it works with Kubernetes, you're pretty much good to go in terms of learning other monitoring tools. It's all the same stuff at the end of the day. The underlying components of what monitoring is doesn't change. The only thing that changes is how the dashboards look.

In this section, you're going to learn about the following:

- The built-in Kubernetes Dashboard
- Cloud-specific monitoring and observability tools
- Grafana/Prometheus
- How to use and set monitoring tools

The Kubernetes Dashboard

The Kubernetes Dashboard is as *native* as it gets in terms of monitoring and observability. Although it's not configured out of the box, it's fairly easy to get configuration across almost any environment. It's the quickest way to see what's happening inside a Kubernetes cluster.

> **Important note**
>
> We're using `minikube` for this because it's straightforward. If you decide to use the Kubernetes Dashboard on another Kubernetes cluster, the visual of the dashboard itself isn't going to be any different. The only difference will be the Kubernetes resources that you see.

First, start `minikube`. If you don't have `minikube` already installed, you can install it here: `https://minikube.sigs.k8s.io/docs/start/`:

```
minikube start
```

```
+ Automatically selected the docker driver
  Starting control plane node minikube in cluster minikube
  Pulling base image ...
  Creating docker container (CPUs=2, Memory=4000MB) ...
  Preparing Kubernetes v1.23.1 on Docker 20.10.12 ...
    ▪ kubelet.housekeeping-interval=5m
    ▪ Generating certificates and keys ...
    ▪ Booting up control plane ...
    ▪ Configuring RBAC rules ...
  Verifying Kubernetes components...
    ▪ Using image gcr.io/k8s-minikube/storage-provisioner:v5
  Enabled addons: default-storageclass, storage-provisioner
  Done! kubectl is now configured to use "minikube" cluster and "default" namespace by default
```

Figure 7.10 – Starting minikube

Next, run the following command to start the dashboard:

```
minikube dashboard -url
```

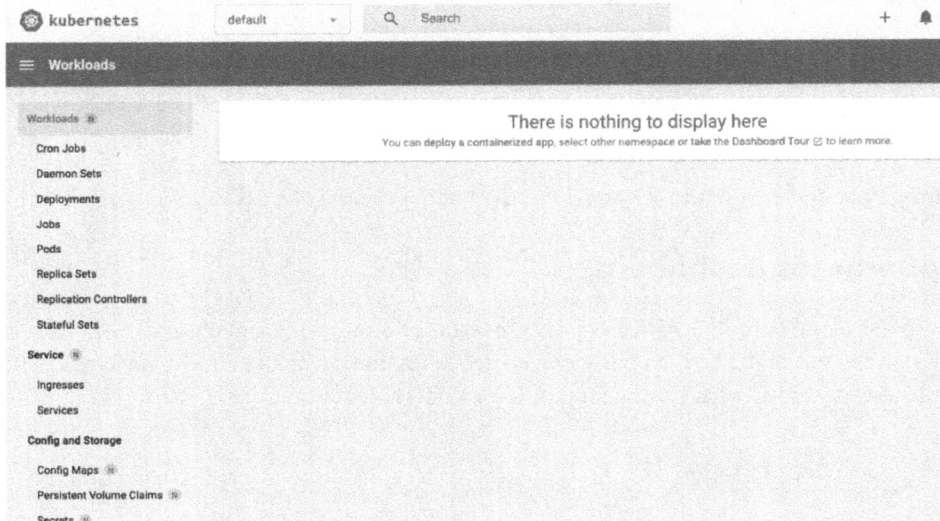

Figure 7.11 – The default Kubernetes Dashboard

At this point, you can see several different pieces of information about your `minikube` cluster, from Pod info to other Kubernetes resources. You can see Pods that are running and healthy, and workloads that may need to be fixed.

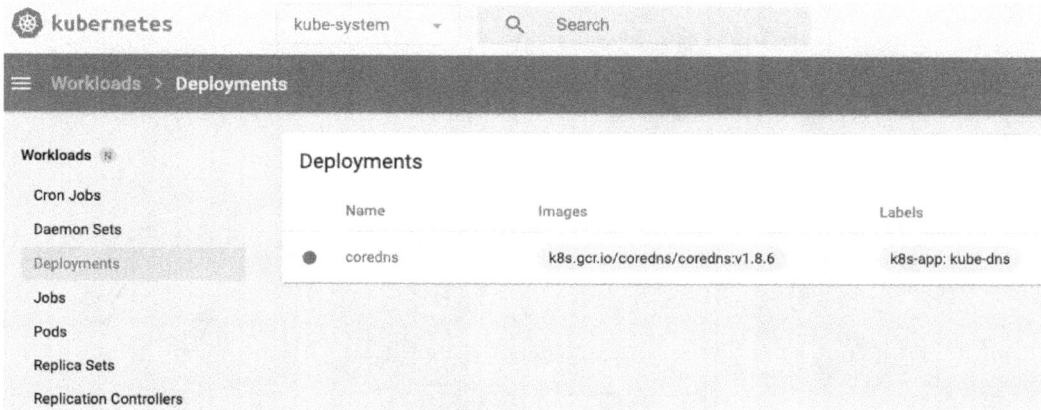

Figure 7.12 – A Deployment example

Next, you can see the overall deployment status.

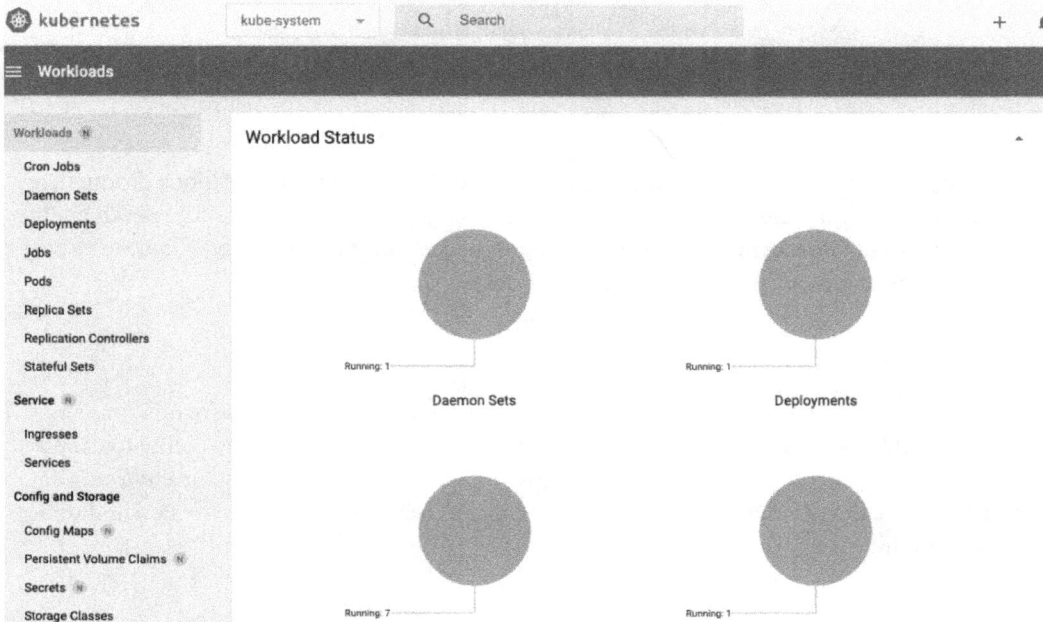

Figure 7.13 – The Pod status

After that, you can dive even deeper to see Pods running in the **Deployments** tab.

aemon Sets

Name	Images	Labels	Pods
● kube-proxy	k8s.gcr.io/kube-proxy:v1.23.1	k8s-app: kube-proxy	1 / 1

eployments

Name	Images	Labels	Pods
● coredns	k8s.gcr.io/coredns/coredns:v1.8.6	k8s-app: kube-dns	1 / 1

ods

Name	Images	Labels	Node	Status	Restarts	CPU Usage (cores)
● coredns-64897985d-v9j6b	k8s.gcr.io/coredns/cored ns:v1.8.6	k8s-app: kube-dns pod-template-hash: 6489 7985d	minikube	Running	0	-
● kube-proxy-lsk69	k8s.gcr.io/kube-proxy:v1. 23.1	controller-revision-hash: 8 485885f8b k8s-app: kube-proxy pod-template-generatio n: 1	minikube	Running	0	-
● storage-provisioner	gcr.io/k8s-minikube/stora ge-provisioner:v5	addonmanager.kubernete s.io/mode: Reconcile integration-test: storage-p rovisioner	minikube	Running	1	-

Figure 7.14 – The Pods running

One thing to point out here is that the Kubernetes Dashboard is almost never used for a production-level scenario. It's typically used to look at some information quickly if needed. For true observability and alerting in an environment, one of the more appropriate (production-ready) monitoring and observability tools is typically used, which you'll see next.

Azure Monitor

If you strictly have Azure workloads or even workloads outside of Azure and you're utilizing Azure Arc (like on-premises), Azure Monitor is a great built-in solution. You have the ability to capture logs and metrics, create alerts, and see in real time what's happening inside your environment. For example, you can view the CPU and memory usage of a cluster, along with the Pod and other Kubernetes resource data.

In *Chapter 2*, you learned how to create an AKS cluster with Terraform. You can utilize that same code for this section. For a quicker reference, here is the link: `https://github.com/PacktPublishing/50-Kubernetes-Concepts-Every-DevOps-Engineer-Should-Know/tree/main/Ch2/AKS`.

Once your AKS cluster is configured, log in to the Azure portal and go to **Kubernetes services**. Then, you should see an **Insights** tab under **Monitoring**.

Enable Insights by clicking the blue **Configure azure monitor** button.

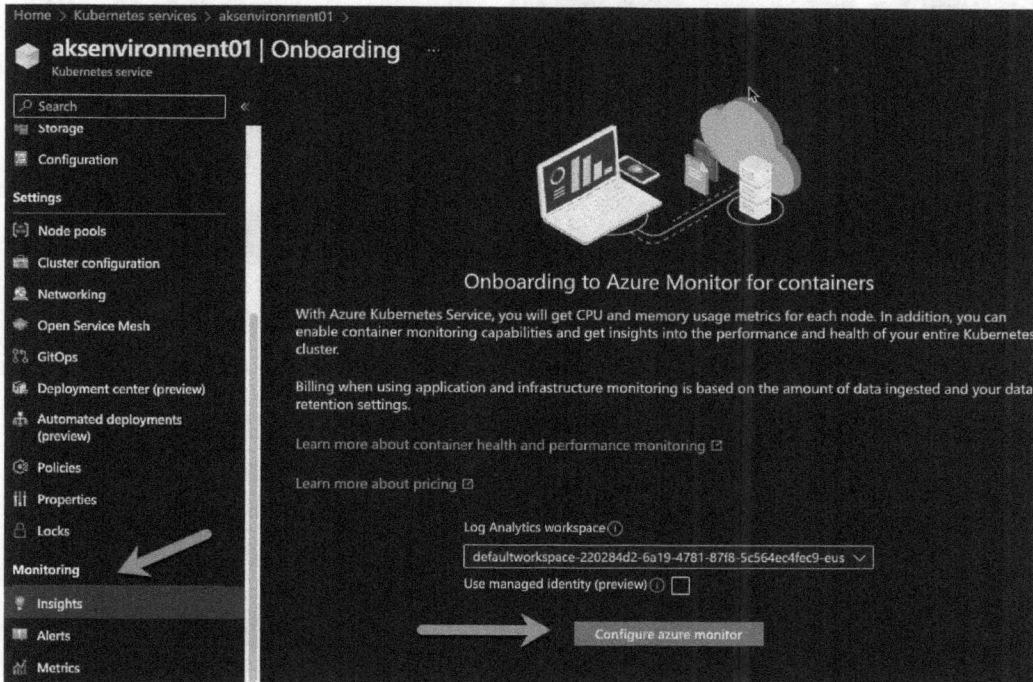

Figure 7.15 – Azure Insights

Azure Insights gives you the ability to monitor everything in your AKS cluster from the entire environment, to the nodes, all the way down to the Pods and containers.

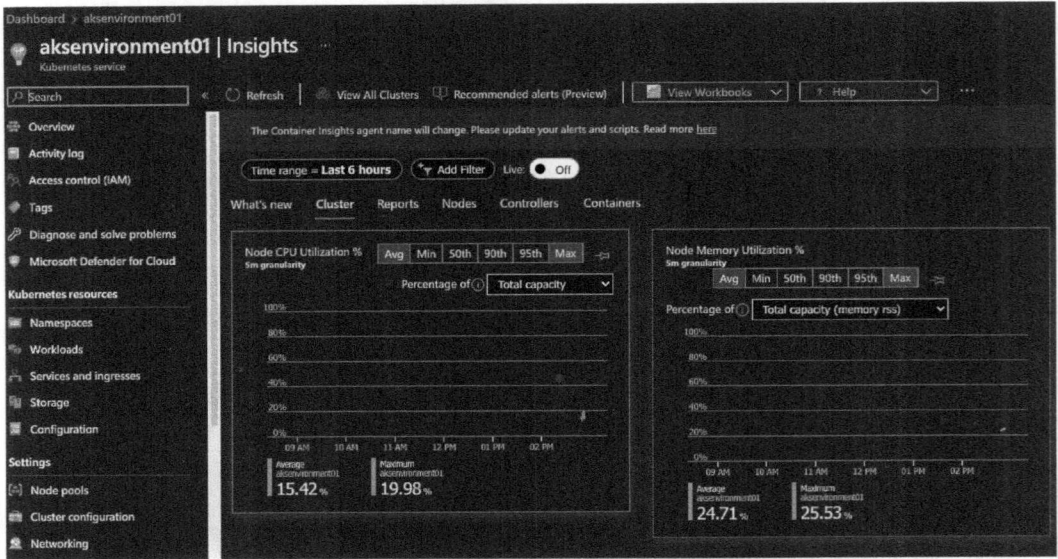

Figure 7.16 – Insights data

For example, by diving into **Containers** (Pods), you can see the status, utilization, and uptime.

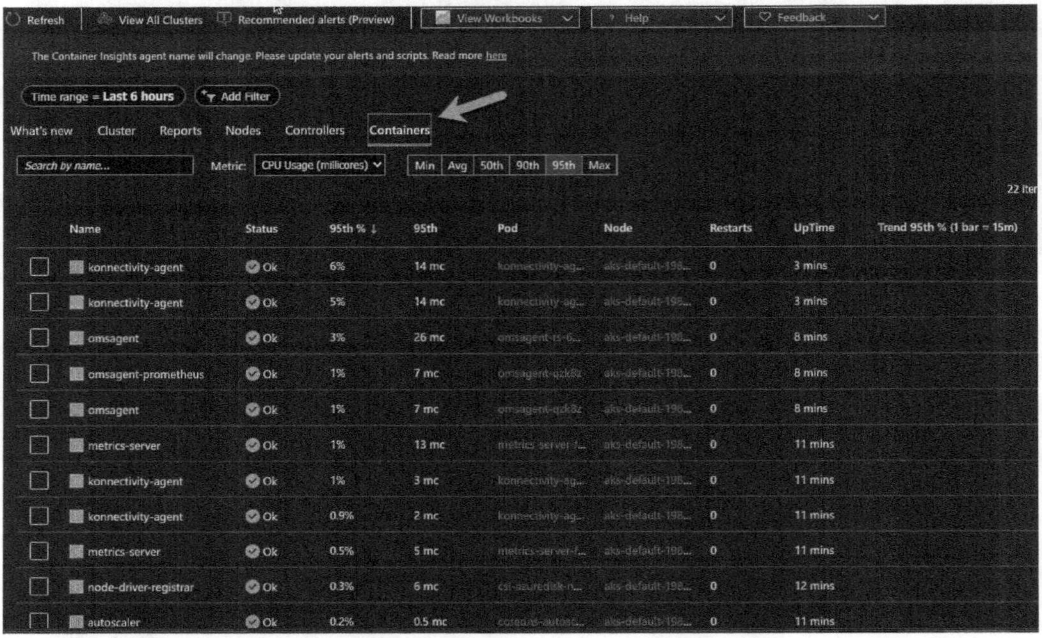

Figure 7.17 – The Container data

Within **Nodes**, you can see the specific Pods running on each worker node, including the health of the Pod.

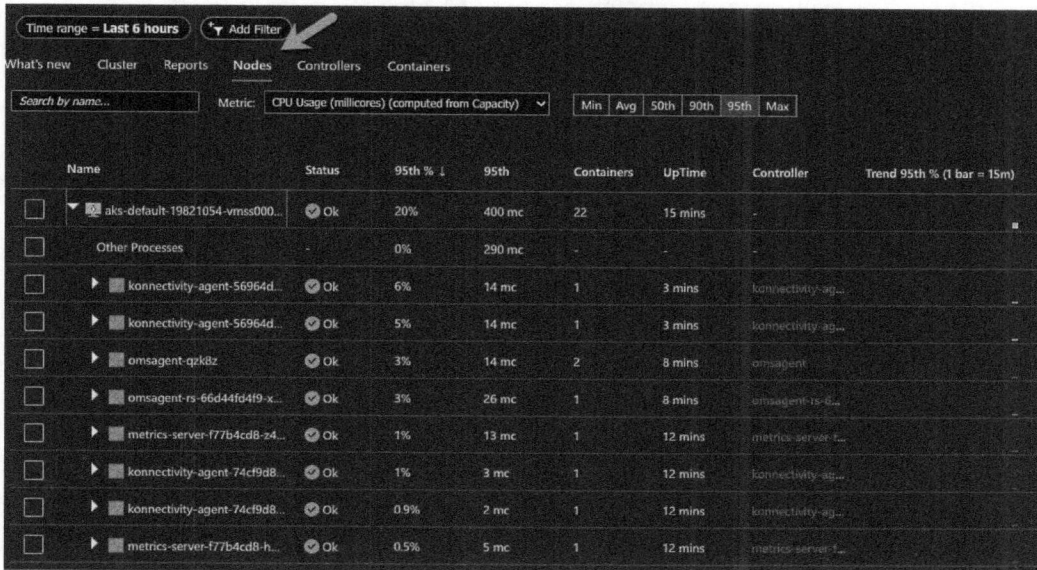

	Name	Status	95th % ↓	95th	Containers	UpTime	Controller	Trend 95th % (1 bar = 15m)
☐	▼ 🖳 aks-default-19821054-vmss000...	✅ Ok	20%	400 mc	22	15 mins	-	
☐	Other Processes	-	0%	290 mc	-	-	-	
☐	▶ 🖳 konnectivity-agent-56964d...	✅ Ok	6%	14 mc	1	3 mins	konnectivity-ag...	
☐	▶ 🖳 konnectivity-agent-56964d...	✅ Ok	5%	14 mc	1	3 mins	konnectivity-ag...	
☐	▶ 🖳 omsagent-qzk8z	✅ Ok	3%	14 mc	2	8 mins	omsagent	
☐	▶ 🖳 omsagent-rs-66d44fd4f9-x...	✅ Ok	3%	26 mc	1	8 mins	omsagent-rs-6...	
☐	▶ 🖳 metrics-server-f77b4cd8-z4...	✅ Ok	1%	13 mc	1	12 mins	metrics-server-f...	
☐	▶ 🖳 konnectivity-agent-74cf9d8...	✅ Ok	1%	3 mc	1	12 mins	konnectivity-ag...	
☐	▶ 🖳 konnectivity-agent-74cf9d8...	✅ Ok	0.9%	2 mc	1	12 mins	konnectivity-ag...	
☐	▶ 🖳 metrics-server-f77b4cd8-h...	✅ Ok	0.5%	5 mc	1	12 mins	metrics-server-f...	

Figure 7.18 – The Node data

Azure Monitor and Insights is a great overall solution for Kubernetes workloads. If you're in the Azure ecosystem, I wouldn't recommend looking at another solution. Stick to what's native.

AWS Container Insights

Container Insights is part of the AWS CloudWatch family and gives you the ability to view containerized workloads for performance and monitoring-related actions. You can create alerts based on Container Insights, along with pull logs and metrics to take action on anything that may occur from an automated and repeatable perspective.

In *Chapter 2*, you learned how to create an EKS cluster with Terraform. You can utilize the same code for this section. For a quicker reference, here is the link: `https://github.com/PacktPublishing/50-Kubernetes-Concepts-Every-DevOps-Engineer-Should-Know/tree/main/Ch2/AWS`.

After you run the EKS Terraform configuration, run the following command to retrieve the Kubernetes configuration (`kubeconfig`) from the EKS cluster:

```
aws eks update-kubeconfig -region region_where_cluster_exists -
name name_of_your_cluster
```

To confirm that your current context is set, run the following command and you should see a similar output:

```
kubectl get nodes
NAME                              STATUS    ROLES     AGE
    VERSION
ip-192-168-16-238.ec2.internal    Ready     <none>    18m
    v1.23.9-eks-ba74326
```

Next, configure AWS Container Insights for your cluster:

```
ClusterName= name_of_your_cluster
RegionName= region_where_cluster_exists
FluentBitHttpPort='2020'
FluentBitReadFromHead='Off'
[[ ${FluentBitReadFromHead} = 'On' ]] &&
FluentBitReadFromTail='Off'|| FluentBitReadFromTail='On'
[[ -z ${FluentBitHttpPort} ]] && FluentBitHttpServer='Off' ||
FluentBitHttpServer='On'
curl https://raw.githubusercontent.com/aws-samples/
amazon-cloudwatch-container-insights/latest/k8s-
deployment-manifest-templates/deployment-mode/
daemonset/container-insights-monitoring/quickstart/
cwagent-fluent-bit-quickstart.yaml | sed 's/{{cluster_
name}}/'${ClusterName}'/;s/{{region_name}}/'${RegionName}'/;s/
{{http_server_toggle}}/"'${FluentBitHttpServer}'"/;s/
{{http_server_port}}/"'${FluentBitHttpPort}'"/;s/{{read_
from_head}}/"'${FluentBitReadFromHead}'"/;s/{{read_from_
tail}}/"'${FluentBitReadFromTail}'"/' | kubectl apply -f -
```

After the preceding code runs, you'll see an output similar to the terminal output pasted here:

```
namespace/amazon-cloudwatch created
serviceaccount/cloudwatch-agent created
clusterrole.rbac.authorization.k8s.io/cloudwatch-agent-role
created
clusterrolebinding.rbac.authorization.k8s.io/cloudwatch-agent-
role-binding created
configmap/cwagentconfig created
daemonset.apps/cloudwatch-agent created
configmap/fluent-bit-cluster-info created
```

```
serviceaccount/fluent-bit created
clusterrole.rbac.authorization.k8s.io/fluent-bit-role created
clusterrolebinding.rbac.authorization.k8s.io/fluent-bit-role-
binding created
configmap/fluent-bit-config created
daemonset.apps/fluent-bit created
```

At this point, if you log in to AWS and go to **CloudWatch | Container Insights**, you can see that Container Insights is properly configured.

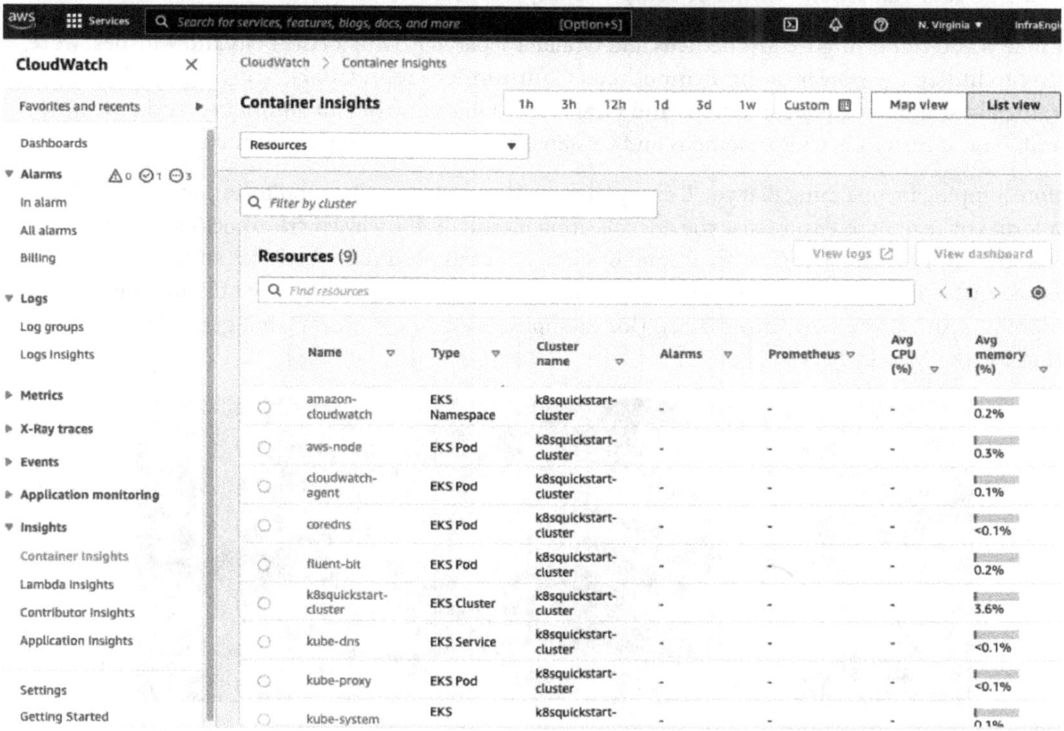

Figure 7.19 – The Container Insights output

Next, we'll dive into a very popular stack in the Kubernetes space – Grafana and Prometheus.

Grafana/Prometheus

Arguably, the most popular implementation of a monitoring/observability scenario for Kubernetes is Grafana and Prometheus. Grafana and Prometheus work outside of Kubernetes environments as well, but they became extremely popular in the Kubernetes ecosystem. In fact, there's even a Prometheus operator for Kubernetes.

Aside from the standard monitoring and observability benefits, engineers really enjoy the combination because it's 100% open source. In Grafana for example, you can create any type of dashboard you want with a little bit of code and it's all free. Grafana and Prometheus can also run anywhere. The stack can run inside your Kubernetes cluster or completely separate on its own servers.

Although you can configure Prometheus and Grafana separately with all the bells and whistles, we're going to utilize the power of the **Prometheus Community Helm Chart**. The reason why is that it radically simplifies the Prometheus and Grafana installation from an automated and repeatable standpoint. It installs both Prometheus and Grafana, along with setting up dashboards for us.

Before jumping in, one thing that you'll always need to do no matter what monitoring and observability platform you're on is to ensure that you are collecting metrics in the way you're expecting. For example, the Kubernetes Metrics Server or an adapter of sorts. For example, Prometheus has an adapter that can be used instead of the Metrics Server. You can also go straight to the source by utilizing the metrics endpoint from `/metrics/resource` (for example, `/metrics/pods`), but generally, engineers opt to use the Metrics Server.

```
NAME                                         READY   STATUS    RESTARTS   AGE
pod/azure-ip-masq-agent-5x7kp                1/1     Running   0          20m
pod/cloud-node-manager-vk5dp                 1/1     Running   0          20m
pod/coredns-autoscaler-5589fb5654-vjzxw      1/1     Running   0          21m
pod/coredns-b4854dd98-nkhcn                  1/1     Running   0          21m
pod/coredns-b4854dd98-qsqfl                  1/1     Running   0          19m
pod/csi-azuredisk-node-h6fwq                 3/3     Running   0          20m
pod/csi-azurefile-node-w9bjk                 3/3     Running   0          20m
pod/konnectivity-agent-5bc84fc8b7-p4mbz      1/1     Running   0          21m
pod/konnectivity-agent-5bc84fc8b7-zn54t      1/1     Running   0          21m
pod/kube-proxy-dfmnc                         1/1     Running   0          20m
pod/metrics-server-f77b4cd8-f42lk            1/1     Running   0          21m
pod/metrics-server-f77b4cd8-m6b8n            1/1     Running   0          21m
```

Figure 7.20 – The metrics Pods

If you don't expose the metrics endpoint, Kubernetes won't allow the system to consume said metrics. In terms of enabling the Metrics Server, it all depends on where you're running Kubernetes. For example, in AKS, it's automatically exposed for you. If you don't see the metrics Pods in the `kube-system` namespace for your Kubernetes cluster (depending on what environment you deployed Kubernetes in), check the documentation for that type of Kubernetes environment to see how you can enable the metrics endpoint.

First, add `helm repo` for `prometheus-community`:

```
helm repo add prometheus-community https://prometheus-
community.github.io/helm-charts
```

Next, ensure that the repo is up to date:

```
helm repo update
```

For the last step, install the Helm chart in the `monitoring` namespace:

```
helm install prometheus prometheus-community/kube-prometheus-
stack -namespace monitoring -create-namespace
```

Once installed, you should see several Kubernetes resources created in the `monitoring` namespace. To access Grafana, you can use port forwarding:

```
kubectl -namespace monitoring port-forward svc/prometheus-
grafana :80
```

The default username/password for Grafana is `admin/prom-operator`.

After logging in to Grafana, check out the Pods in the dashboard for the `kube-system` namespace. You can see that metrics are being ingested by Prometheus and pushed to Grafana from all namespaces.

To see metrics, go to **Dashboards | Browse**:

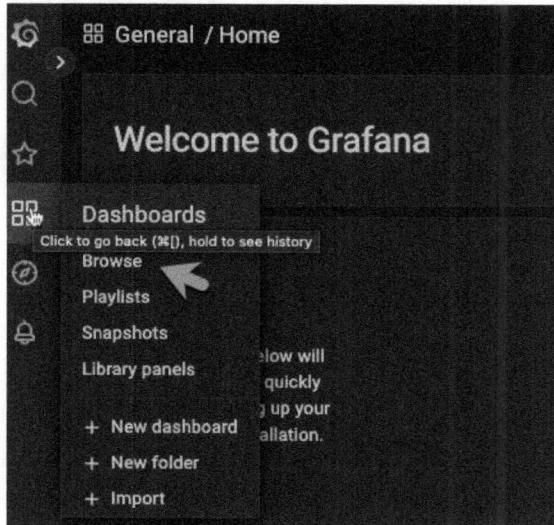

Figure 7.21 – Browsing the dashboards

Click the **Kubernetes / Compute Resources / Namespace (Pods)** option:

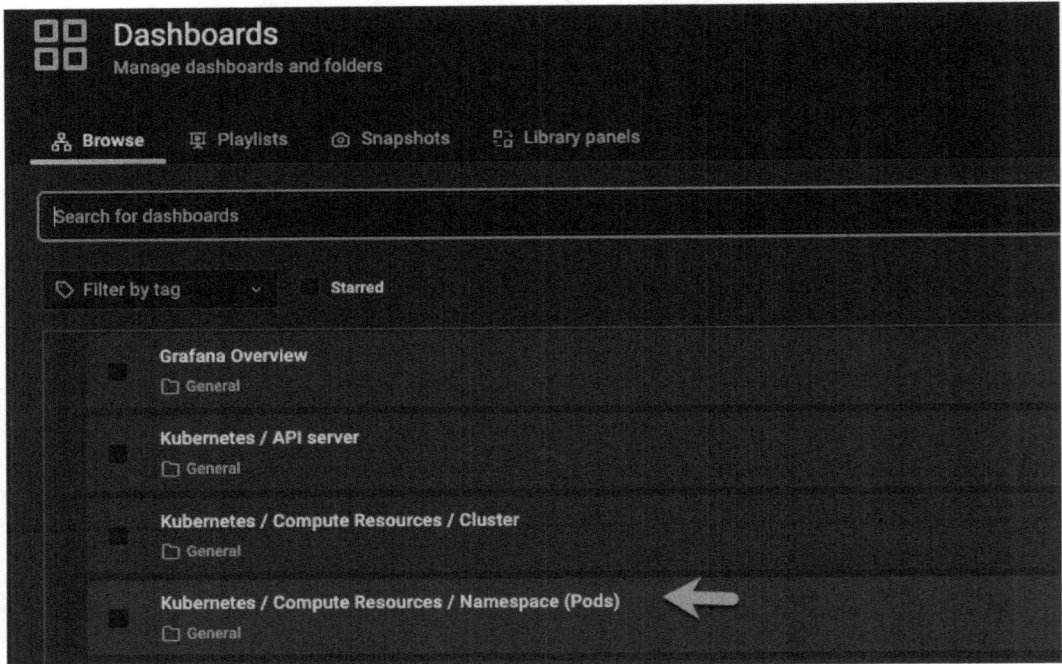

Figure 7.22 – The Pods dashboard

Change the namespace to a namespace that has Pods already, such as kube-system, and you can see the Pod metrics in the following screenshot:

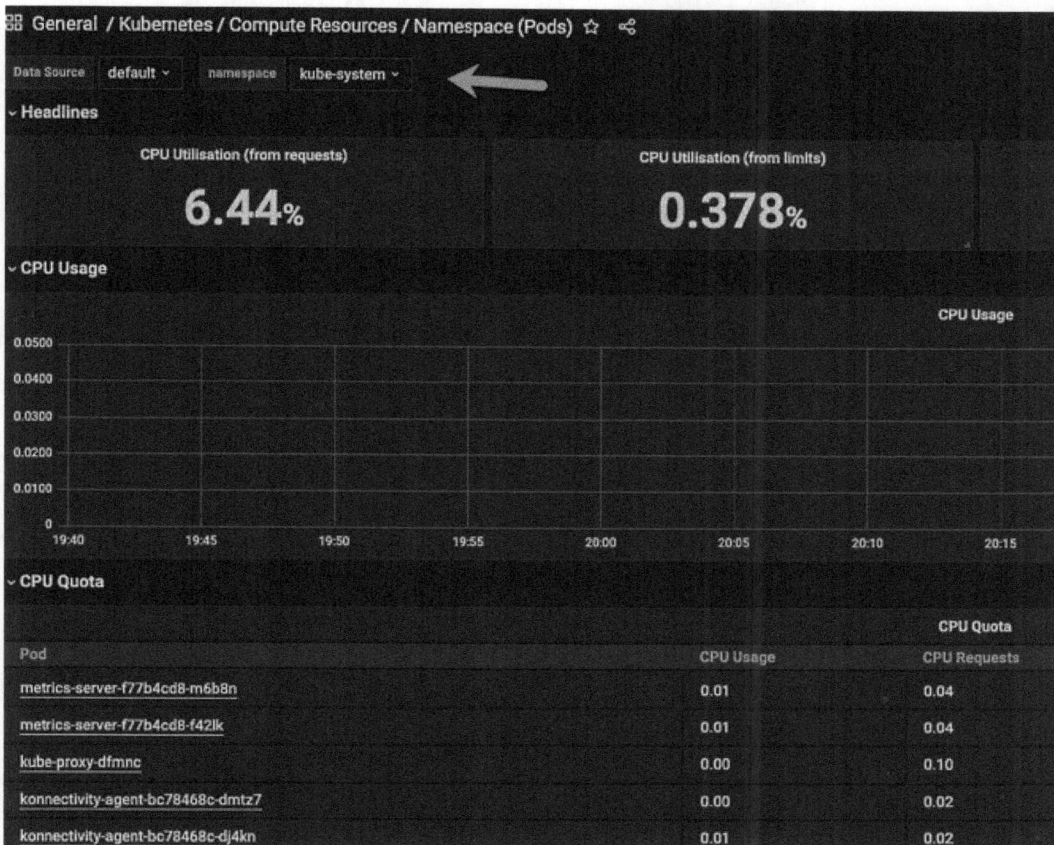

Figure 7.23 – The namespace selection

Prometheus/Grafana is a powerful combination that allows you to stay vendor neutral and get everything you need as an open source option.

Observability practices

Now, let's define what observability truly is by looking at logs, traces, and metrics. When you use tools such as Prometheus, you're doing a *piece* of observability. When you use other tools such as Logz.io or another log aggregator, you're using another piece of observability.

Logging

Logging is aggregating and storing logged event messages written by programs and systems. As you can imagine, depending on how verbose the logs are set in an application, there will be a lot of events. A sysadmin's favorite tool is a log because it literally shows everything and anything that could happen from an event's perspective. However, it's not efficient to simply comb through all of it with your eyes. Instead, using observability practices, you can send the logs to a log aggregator and ensure that a specific type of log that occurs can trigger an alert or some type of automation to go in and fix the issue.

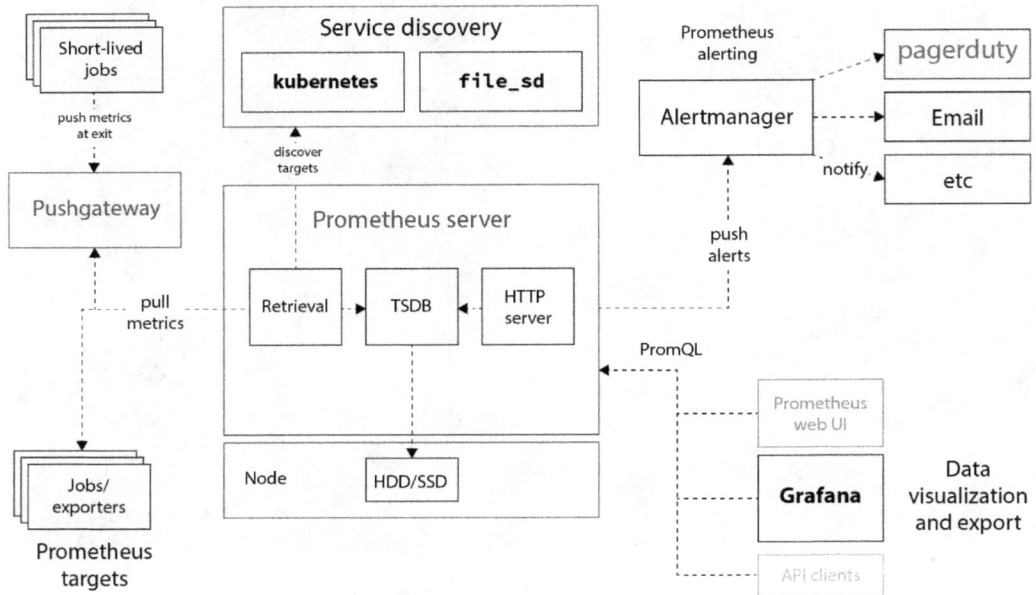

Figure 7.24 – Logging service discovery

There are a few logging practices when it comes to containers:

- **Application forwarding**: Sending logs directly via the app. For example, maybe you have some code inside of your application using a Prometheus library that collects the logs, metrics, and traces, and sends it to whatever backend logging platform you're using.

- **Sidecar**: Using a sidecar container to manage logs for an app. For example, you can containerize some logging systems to run as a secondary/sidecar container inside of your Pod(s). The sidecar container's job is to do one thing; retrieve and send logs about what's happening on the Pod.

- **Node agent forward**: Run a Pod on each worker node that forwards all container logs to the backend.

Metrics

Metrics are about collecting time series data, which is used to predict expected ranges and forecast values, showing it in dashboards (such as Grafana or another UI-centric dashboard), and alerting on it. Metric endpoints will give a bunch of information that you can act upon. From a pure Kubernetes perspective, the metrics endpoint collects Kubernetes resource data from the kubelet that's running on each worker node and exposes it to the API server through the Metrics API.

As mentioned in this chapter, there's a metrics endpoint that runs as a Pod. Depending on the type of Kubernetes cluster you're running, the Pod could either be enabled by default or it may be something that you have to turn on.

For example, in an AKS cluster, the metrics Pod is running, which means all of the Kubernetes resources have a metrics endpoint that can be consumed.

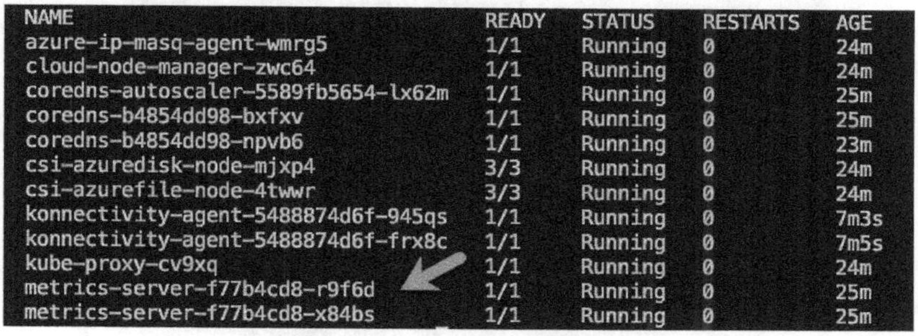

```
NAME                                   READY   STATUS    RESTARTS   AGE
azure-ip-masq-agent-wmrg5              1/1     Running   0          24m
cloud-node-manager-zwc64               1/1     Running   0          24m
coredns-autoscaler-5589fb5654-lx62m    1/1     Running   0          25m
coredns-b4854dd98-bxfxv                1/1     Running   0          25m
coredns-b4854dd98-npvb6                1/1     Running   0          23m
csi-azuredisk-node-mjxp4               3/3     Running   0          24m
csi-azurefile-node-4twwr               3/3     Running   0          24m
konnectivity-agent-5488874d6f-945qs    1/1     Running   0          7m3s
konnectivity-agent-5488874d6f-frx8c    1/1     Running   0          7m5s
kube-proxy-cv9xq                       1/1     Running   0          24m
metrics-server-f77b4cd8-r9f6d          1/1     Running   0          25m
metrics-server-f77b4cd8-x84bs          1/1     Running   0          25m
```

Figure 7.25 – The metrics Pod

For another type of Kubernetes cluster, such as something running on Kubeadm, you would have to enable the metrics endpoint by deploying the Pod. You can do that by deploying the Kubernetes manifest in the `kubernetes-sigs` repo on GitHub:

```
kubectl apply -f https://github.com/kubernetes-sigs/metrics-
server/releases/latest/download/components.yaml
```

However, that's not all. Because a cluster configuration such as Kubeadm has node IPs that aren't part of the certificate SAN on the cluster, the metrics endpoint will fail due to a TLS connection error.

To get around this, you have to add the following line to a few configurations:

```
serverTLSBootstrap: true
```

There are two places you need to add it to.

First, the Kubeadm config. You can edit it by running `kubectl edit cm -n kube-system kubeadm-config` and then add in the `serverTLSBootstrap: true` line.

```yaml
apiVersion: v1
data:
  ClusterConfiguration: |
    apiServer:
      extraArgs:
        authorization-mode: Node,RBAC
      timeoutForControlPlane: 4m0s
    apiVersion: kubeadm.k8s.io/v1beta3
    certificatesDir: /etc/kubernetes/pki
    serverTLSBootstrap: true          <---
    clusterName: kubernetes
    controlPlaneEndpoint: 192.168.1.61:6443
    controllerManager: {}
    dns: {}
    etcd:
      local:
        dataDir: /var/lib/etcd
    imageRepository: registry.k8s.io
    kind: ClusterConfiguration
    kubernetesVersion: v1.25.2
    networking:
      dnsDomain: cluster.local
      podSubnet: 172.17.0.0/16
      serviceSubnet: 10.96.0.0/12
    scheduler: {}
kind: ConfigMap
metadata:
  creationTimestamp: "2022-10-12T21:13:48Z"
  name: kubeadm-config
  namespace: kube-system
  resourceVersion: "396121"
  uid: d359b06d-a128-499e-b2ee-ef29bd5c1e28
~
"/tmp/kubectl-edit-3472111723.yaml" 36L, 1052B
```

Figure 7.26 – The kubeadm config

Next, you'll have to update the kubelet on each node (all Control Planes and worker nodes) with the same line. To edit the kubelet on each node, you can run the following command and add in the configuration:

```
sudo vim /var/lib/kubelet/config.yaml
```

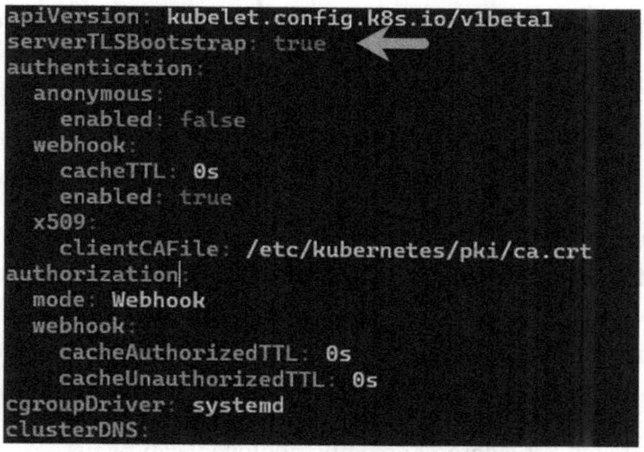

```
apiVersion: kubelet.config.k8s.io/v1beta1
serverTLSBootstrap: true      ←
authentication:
  anonymous:
    enabled: false
  webhook:
    cacheTTL: 0s
    enabled: true
  x509:
    clientCAFile: /etc/kubernetes/pki/ca.crt
authorization:
  mode: Webhook
  webhook:
    cacheAuthorizedTTL: 0s
    cacheUnauthorizedTTL: 0s
cgroupDriver: systemd
clusterDNS:
```

Figure 7.27 – The kubelet config

Traces

Traces are all about telling you the health of an application from an end-to-end perspective.

You can think of a trace as the *path* or *journey* of a request as it goes through the system. For example, when you go to www.google.com, although it happens extremely fast, there's a bunch of work that's happening underneath the hood. At a high level, the GET request that you're creating to reach www.google.com is going through the frontend, then probably some middleware, then to the backend. When you Google something such as top ten places to go in the summertime, there are several requests that are occurring to retrieve that information from the backend database.

The journey from when you perform a Google search request to when the information is portrayed to you – that journey is what a trace is.

Because it's a long journey, although only seconds to us humans, it can give us a lot of information from an engineering perspective on how an application is performing. We can then take action on that performance concern from a repeatable methodology instead of fixing the issue manually, or from a troubleshooting perspective. If you're looking at a trace and realize that the *journey* stopped or was held up once it hit the backend, you now know where to start troubleshooting.

Monitoring Kubernetes resources

In the previous section, you learned all about monitoring from an overall observability perspective, in particular setting up certain tools and ensuring that they work for you. Now it's time to go underneath the Kubernetes hood and begin to think about what can be monitored from a resource perspective. Remember, a Kubernetes resource (sometimes called an object) can be anything, from Services, to Ingress controllers, to Pods. Because of that, there's a lot to monitor.

Think about it from this perspective. You're running a Pod that's running a container inside of the Pod. The Pod itself is running great. The container image works, with no CPU or memory issues, and all of the events state that the Pod is up and running successfully. However, there's a problem – the binary (the app entry point) running inside of the container may be down, or not working as expected. Because of this, you need a way to truly see even underneath the hood of a Pod itself!

As you've learned throughout this book, it doesn't really matter where you're running Kubernetes. The core components of how it runs and how you would interact with it are the same. That's no different for monitoring. Because of that, this section of the chapter will show monitoring in AKS. However, as you'll quickly see, it doesn't matter whether these Pods are running in AKS or not. They would be looked at (monitored) the same way even if you use a different monitoring system.

The code in this section, along with the demo app being deployed, can be used on any Kubernetes cluster.

Monitoring Pods

Inside a Pod is either one or more containers. Whether it's one container or multiple containers, the containers are what's actually running an application. Perhaps it's a core app, a logging software, or even something such as HashiCorp Vault or a service mesh proxy. These containers that are beside the main app are called sidecar containers. Because there are multiple containers running inside a Pod, you must ensure that each container is actually up and running as expected. Otherwise, the Pod itself may be running properly, and the main app may even be running properly, but the full workload, such as the sidecar containers, may not be.

First, ensure that the HashiCorp Consul Helm chart exists:

```
helm repo add hashicorp https://helm.releases.hashicorp.com
```

Next, create a new namespace called `consul`:

```
kubectl create namespace consul
```

Once the `consul` namespace exists, deploy Consul to Kubernetes inside of the `consul` namespace:

```
helm upgrade -install -n consul consul hashicorp/consul -wait
-f - <<EOF
global:
  name: consul
server:
```

```
  replicas: 1
  bootstrapExpect: 1
connectInject:
  enabled: true
EOF
```

The last step is to deploy the demo app and ensure that the annotation for injecting `consul` as a sidecar exists:

```
Curl -sL https://run.linkerd.io/emojivoto.yml \
  | sed 's|        metadata: |        metadata:\n        annotations:\n
          consul.hashicorp.com/connect-inject" "tr"e'|' \
  | se' 's|targetPort: 8080|targetPort: 2000'|' \
  | kubectl apply -f -
```

After deploying the app, you should see an output similar to the following screenshot:

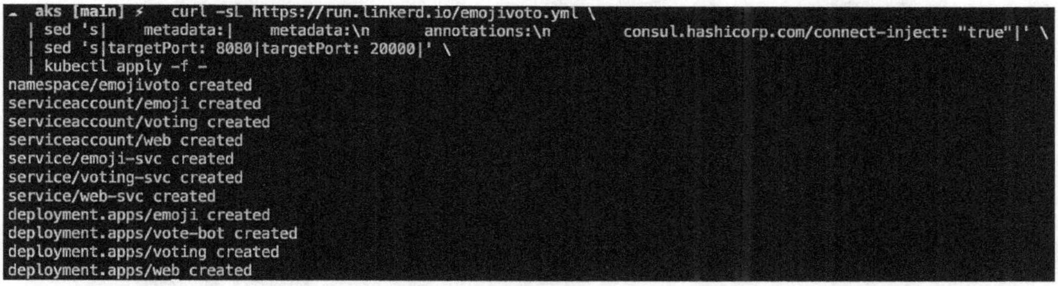

Figure 7.28 – Linkerd deployment

Log in to the Azure portal, go to your AKS cluster, and turn on Azure Insights if it's not already on:

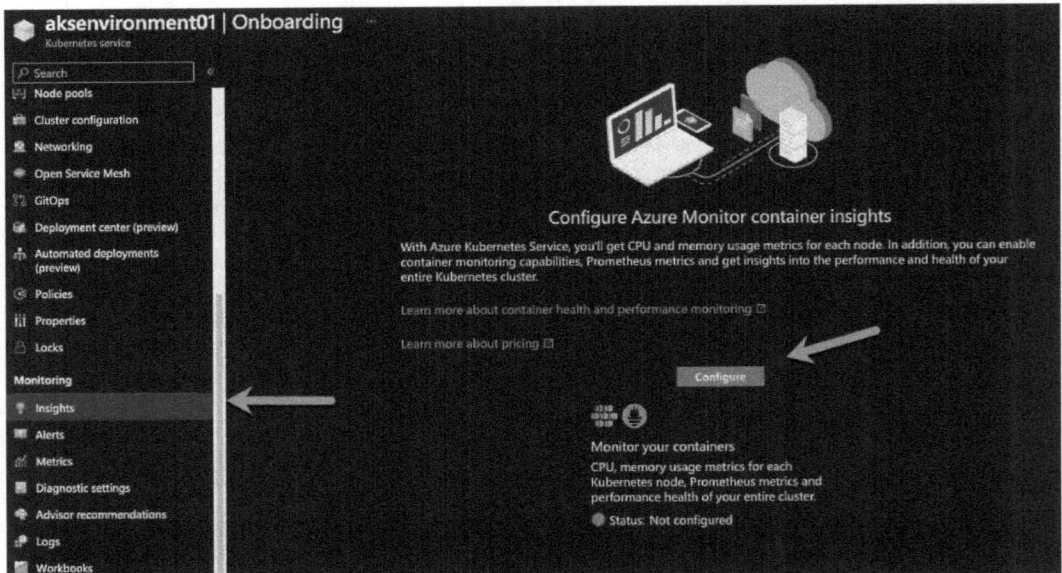

Figure 7.29 – Enabling container insights

Once **Insights** is enabled, you should be able to see several resources available. Click on the **Controllers** button.

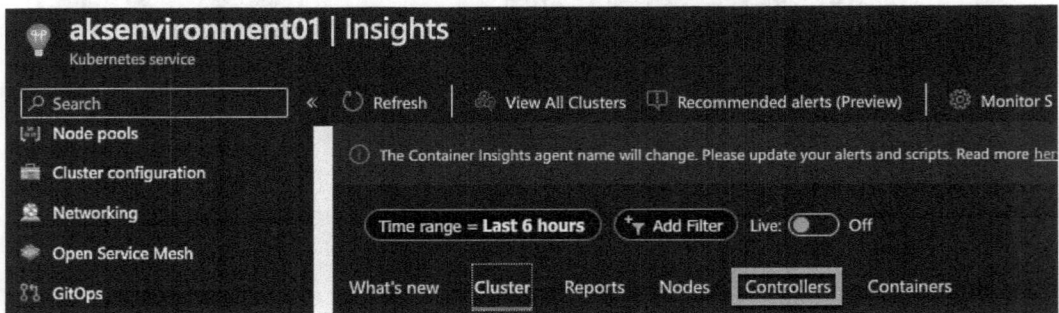

Figure 7.30 – The Controllers dashboard

Looking at the **Controllers** dashboard, you can see all the Kubernetes resources running along with the status, uptime, and how many containers exist in each resource.

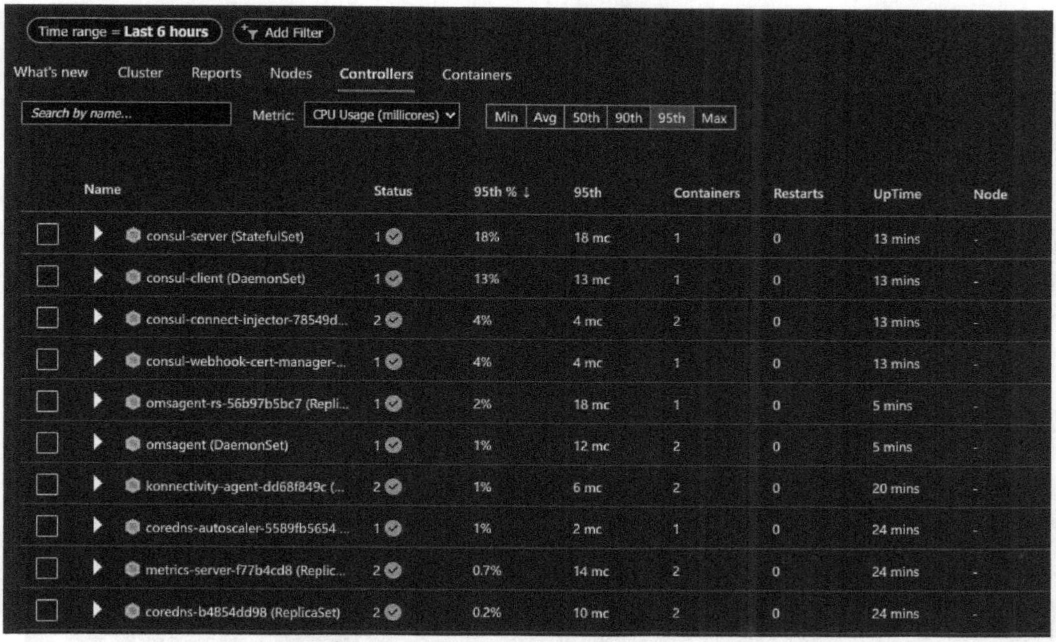

Figure 7.31 – The Kubernetes resources running

Drilling in a bit deeper, you can see that for each resource with more than one Pod, you're able to see the different containers available.

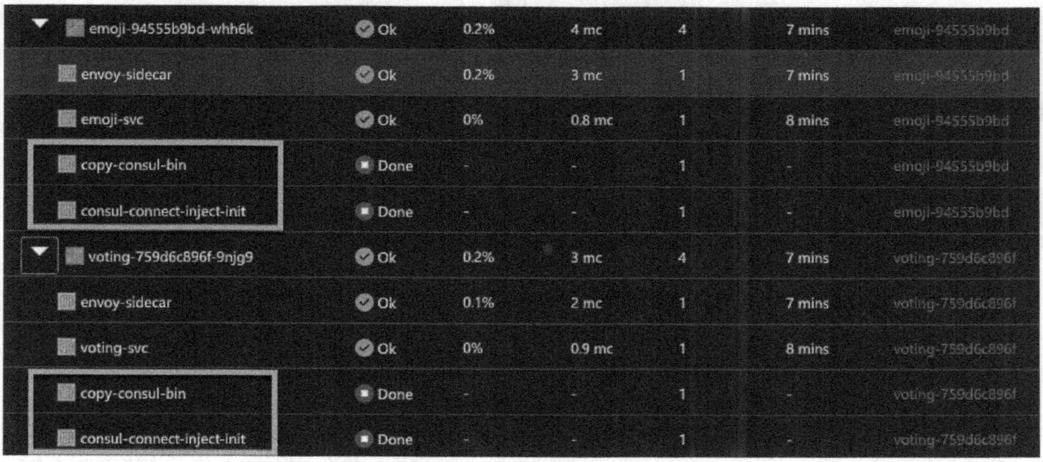

Figure 7.32 – The resources in Pods

But as always, things may go wrong. You can see in the following screenshot that there's a Kubernetes resource running, but some of the containers aren't running as expected:

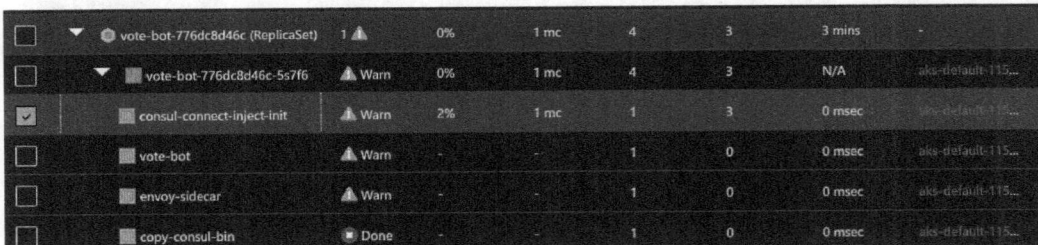

Figure 7.33 – The warning resource

As you dive into it a bit deeper, you can see the status of the container is waiting to be created.

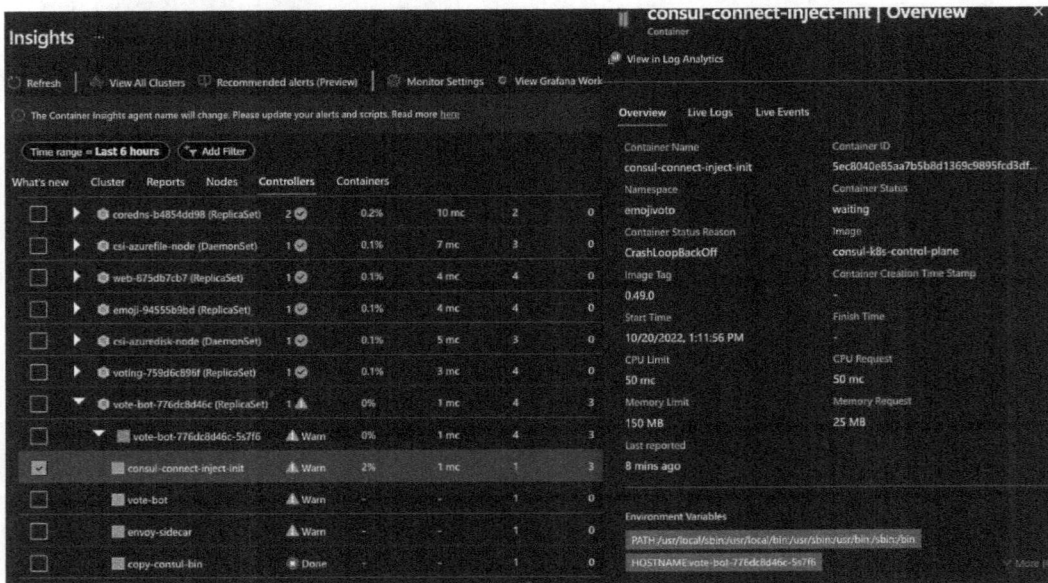

Figure 7.34 – The warning explanation

So, even though the Pod may be up and running, as in the application running inside of the container, other sidecar containers may not be. On the outside looking in, the app is up so it appears that everything is working as expected. However, after giving it a closer look, you can see that it's not. This is the big difference between monitoring an app running on a server and a Pod. Within a Pod, there may be more than one binary to worry about.

Summary

This chapter covered a lot. It's roughly 35 pages, and the thing is, these topics can be two or three books in themselves. Because of that, not everything was covered at the specific depth that's most likely needed. However, the good news is that you now have a solid understanding of how to start thinking about implementing these platforms, technologies, and methodologies in production.

We went over quite a few topics in this chapter, covering what monitoring is, what observability is, and the overall differences between the two. You then dove into the specific tools and platforms available to make monitoring and observability come to life in your Kubernetes environment.

In the next and final chapter, you'll learn about security from a Kubernetes perspective.

Further reading

- *Hands-On Kubernetes on Azure – Second Edition* by Nills Franssens, Shivakumar Gopalakrishnan, and Gunther Lenz: https://www.packtpub.com/product/hands-on-kubernetes-on-azure-second-edition/9781800209671

- *Hands-On Infrastructure Monitoring with Prometheus* by Joel Bastos and Pedro Araújo: https://www.packtpub.com/product/hands-on-infrastructure-monitoring-with-prometheus/9781789612349

Security Reality Check

Security in general, and especially in Kubernetes, is an ironic thing. Everyone knows it's important, yet it's not held to the same necessity as, for example, developers. In fact, if you look at the ratio, there's probably 1 security engineer to 100 developers. Environments aren't secure out of the box, especially when it comes to access control, yet security is arguably one of the most overlooked pieces of Kubernetes. Because of the lack of security awareness around Kubernetes, this chapter is going to focus on a little bit of everything that you should be thinking about when securing a Kubernetes environment.

From a theoretical perspective, you'll be learning how to think about security in Kubernetes. From a hands-on perspective, you'll be learning not only how to implement security practices, but which tools and platforms to use.

When thinking about production, this chapter may very well be the most important one in this entire book. You must walk before you run, and therefore, you must learn how to use Kubernetes in production before you can secure it. The focus of *Chapters 1-7* was to get you to that point. This chapter, however, is all about taking things to the next level, and as with most areas of **Information Technology (IT)**, that *next level* is security.

By the end of this chapter, you'll know which practices to utilize when securing a Kubernetes environment from the cluster itself to the containerized applications running inside of the cluster. You'll also know which tools and platforms to use to get the job done.

In this chapter, we're going to cover the following main topics:

- Out-of-the-box Kubernetes security
- Investigating cluster security
- Understanding **role-based access control (RBAC)**
- Kubernetes resource (object) security
- Kubernetes Secrets

> **Important note**
> Much as with all other engineering-related books and research analyst analyses, the figures/
> percentages used within this chapter are based on various experiences in the field. In this book,
> if there are figures that do not have specific associated data sources, the data is collated from
> the production experiences of the author, Michael Levan.

Technical requirements

For this chapter, as with most of the chapters in this book, you will need a Kubernetes cluster running. Although you can run these tests on something such as Minikube, it's highly recommended to create a Kubeadm cluster or a Kubernetes managed service cluster in the cloud with something such as **Azure Kubernetes Service (AKS)**, **Amazon Elastic Kubernetes Service (Amazon EKS)**, or **Google Kubernetes Engine (GKE)**. The reason why is that you should see what it's truly like from a production perspective to run Kubernetes security tests, which will open your eyes to see how secure (or insecure) it is out of the box and what you can do to mitigate those risks.

If you want to deploy a Kubeadm cluster, check out this Git repo for help:

```
https://github.com/AdminTurnedDevOps/Kubernetes-Quickstart-
Environments/tree/main/Bare-Metal/kubeadm
```

For the overall code used in this chapter, you can find it here:

```
https://github.com/PacktPublishing/50-Kubernetes-Concepts-Every-
DevOps-Engineer-Should-Know/tree/main/Ch8
```

Out-of-the-box Kubernetes security

At this point in time, there are two typical groups of people—those who are so incredibly new to Kubernetes and those who are as close to an *expert* as possible.

With the group that's new to Kubernetes, they're just trying to understand the breakdown of the environment. They aren't even at the stage of thinking about security yet.

With the group that's advanced—yes, they're implementing security practices. The problem is that the advanced group is extremely small compared to the group that's new to Kubernetes.

Then, there are the engineers that are somewhat in between. They aren't super new, but they aren't ridiculously advanced either. This is the group that a lot of engineers fall into, and quite frankly, the group that's somewhat in between is just starting to think about security.

As with most platforms, nothing is 100% secure out of the box. In fact, regardless of how much time you spend to secure an environment, it will never be 100%. The whole goal of security is to mitigate as much risk as possible, but you'll never be able to mitigate 100% of the risk.

From a theoretical perspective, let's talk about a few things around overall security and Kubernetes security.

Security breakdown

Cybersecurity by definition is the protection of systems and networks from system disclosure. This means the protection of anything from the physical server/computer itself to the operating system to any data and metadata on the server/computer or network. If you think about it, that's a lot of information. How many emails do you think get sent through Gmail per day? The specific number for Gmail isn't certain, but for all email providers, the number is collectively 319.6 billion (with a B). Thinking about it from a theoretical but most likely accurate perspective, it's safe to guess that at least 25% of that is Gmail.

The point?

Emails alone contain a massive amount of information, but what about everything else? Information getting sent through networks from one country to another; data on hard drives: there's a lot that falls into the *protection of systems and networks.*

Norton states in a recent blog (`https://us.norton.com/blog/emerging-threats/ cybersecurity-statistics#`) that there are roughly 2,200 cybersecurity attacks per day. To be honest, that seems a bit low. However, even if that number is accurate, that's 800,000 cyberattacks per year. It's certainly no small number.

With that knowledge, as engineers, we must prepare our systems and networks for such types of behavior. As the cloud continues to grow and Kubernetes becomes more mainstream, there will be more attacks directly related to Kubernetes.

As discussed in the opening of this section, the idea of security isn't to stop all risks. The truth is, you'll never be able to stop everything. The security tools, platforms, and engineers that focus on security implementations have one goal in mind—to stop as many security threats as possible. If a system is secure, the operating system may not be. If the operating system is secure, the network may not be. If the network is secure, the applications may not be… and around and around we go. Security is something that can never be 100%, but engineers can take precautions to get as close to 100% as possible.

Thinking about everything in this section, the question comes back to this: *What is security?* In short, it's a method of protecting data.

Kubernetes security

The *State Of Kubernetes* security report from Red Hat (`https://www.redhat.com/en/resources/state-kubernetes-security-report`) highlights security issues directly related to the Kubernetes security landscape:

- 93% of respondents experienced at least 1 security incident in their Kubernetes environments in the last 12 months.

- More than half of the respondents (55%) have had to delay an application rollout because of security concerns.

- Around 70% of security issues in Kubernetes are due to misconfigurations (according to Gartner, it's 99%).

When you look at these statistics from a security report coming right from Red Hat, there's a trend that everyone can easily see—security is a huge issue in the Kubernetes space.

The truth is, as many engineers and executives will attest, security is an absolute mess in the Kubernetes space right now. There's no specific reason why, but there's an educated guess as to why. If you look at the preceding statistics from Red Hat stating that 70% of security issues are due to misconfigurations, that means the primary reason is that engineers are still trying to figure out Kubernetes.

As you've learned about in this book, and as I'm sure you've seen online, almost everyone is still trying to figure out Kubernetes. There's no expert in Kubernetes because the landscape changes every day. There's no end goal to all things Kubernetes because it constantly changes. It's not like a math equation where once you solve it, it's complete. Once you *solve* Kubernetes, 10 more things around Kubernetes would emerge. Because of that, how could a configuration not be misconfigured most of the time? This goes especially for engineers that aren't just focused on Kubernetes, but focused on many areas as well. How can engineers be as close to *experts* as possible within Kubernetes if it's always changing? Misconfigurations are constantly bound to occur.

Because of that, the landscape of Kubernetes security is a mess. In fact, it most likely will be for a long time. It's tough to secure something that's constantly changing.

There's some light at the end of the tunnel, though. As with all platforms and environments, there are best practices that you can follow. Again, thinking about security, what's the goal? To not fix all problems, but to mitigate as many as possible. The purpose of this chapter is to do exactly that: to mitigate as many security risks inside of your Kubernetes environment as possible.

Let's jump in!

Investigating cluster security

Taking Kubernetes out of the equation, let's think about overall infrastructure and/or cloud security. At a high level, you have the network, the servers, the connections to the servers, user access, and ensuring that the applications installed on the servers are secure. In the world of cloud computing, you don't have to worry about the physical security aspect. But if your clusters are in a data center, you do have to think about physical security. Locks on the data center rack cages ensure that no one can plug in any old USB key and that no one can literally take a server out of the rack and walk away with it.

Server security is a combination of what's running inside and on the server—the applications running, programs that are being executed, and the overall operating system itself. Let's say, for example, you're running an older version of Ubuntu. Chances are you should absolutely check and confirm that there are no security holes. That's still very important for any Kubernetes cluster running on Ubuntu. However, Kubernetes has its own set of standards.

From a networking perspective, security still holds true in Kubernetes as well as in any other environment. If you have a frontend or backend Kubernetes service that's accepting traffic from anywhere, that essentially means you have a blanket open firewall. If you're not encrypting Pod-to-Pod and/or service-to-service communication with something such as a service mesh or a security-centric **Container Network Interface (CNI)**, you could open yourself up to more risks.

For example, Kubernetes by definition is an API. As with all APIs, there can be security risks. That means one of the biggest security focus points is to ensure that the Kubernetes API version that you're currently on doesn't have a major security risk as that could literally take down your entire environment.

A big portion of Kubernetes security is benchmarks and other automated testing, which you'll learn about in this section.

Cluster hardening and benchmarks

The **Center for Internet Security (CIS)** has been the de facto standard of hardening systems for years. CIS benchmarks are a set of globally identified standards and best practices when it comes to helping engineers set up their security defenses. Whether it's in the cloud, on-prem, or a specific application/tool, there's a best practice for it, and that's exactly what CIS helps you figure out.

Because CIS is essentially a list of best practices, you have to imagine that there are thousands of different best practices spread across platforms and environments. If you think about a Linux distro, such as Ubuntu, there are specific best practices for that distro alone. If you think about across an entire platform such as **Amazon Web Services (AWS)**, there are even more best practices.

As you look at CIS in general, you'll see that there are a ton of prepopulated CIS environments. For example, in AWS, there are CIS-hardened **Amazon Machine Images** (**AMIs**):

Figure 8.1 – Hardened AMI

In other clouds, such as **Google Cloud Platform** (**GCP**) or Azure, there's the same thing. Even on phones such as an iPhone, there are CIS benchmarks:

Figure 8.2 – iOS hardening

CIS can literally be an entire book in itself, so here's the takeaway—CIS benchmarks are a list of best practices and standards to follow from a security perspective across systems, platforms, apps, and environments.

Because of the popularity of Kubernetes, in 2017, CIS worked with the community to create a benchmark specifically for Kubernetes:

Figure 8.3 – Securing Kubernetes

There are even CIS benchmarks for specific Kubernetes environments, such as GKE.

As you go through this chapter, and as you go through your Kubernetes security journey in general, a lot of tools and platforms you'll see that do things such as container image scanning and cluster scanning use CIS benchmarks. Platforms such as Checkov, `kube-bench`, Kubescape, and a few of the other popular tools in the security space all scan against CIS and the **National Vulnerability Database (NVD)**.

You can download the latest Kubernetes CIS benchmark for free. You just need to put in your name and email at `https://www.cisecurity.org/benchmark/kubernetes`.

Going over the Kubernetes CIS benchmark

The CIS benchmarks in Kubernetes is a huge PDF that you can download and go through to ensure that how you're implementing a Kubernetes environment is up to the best standards and best practices possible for the Kubernetes API version that you're running.

Let's learn how to download the PDF for the Kubernetes CIS benchmark. Follow these steps:

1. Go to this link and fill in your information: `https://www.cisecurity.org/benchmark/kubernetes`.

2. After the information is filled in, you should get an email to download the PDFs. There are going to be a lot, so search for `Kubernetes`. You should then see all the Kubernetes benchmarks.

3. Choose the first one, which at the time of writing this, is for Kubernetes API version 1.23, and click the orange **Download PDF** button:

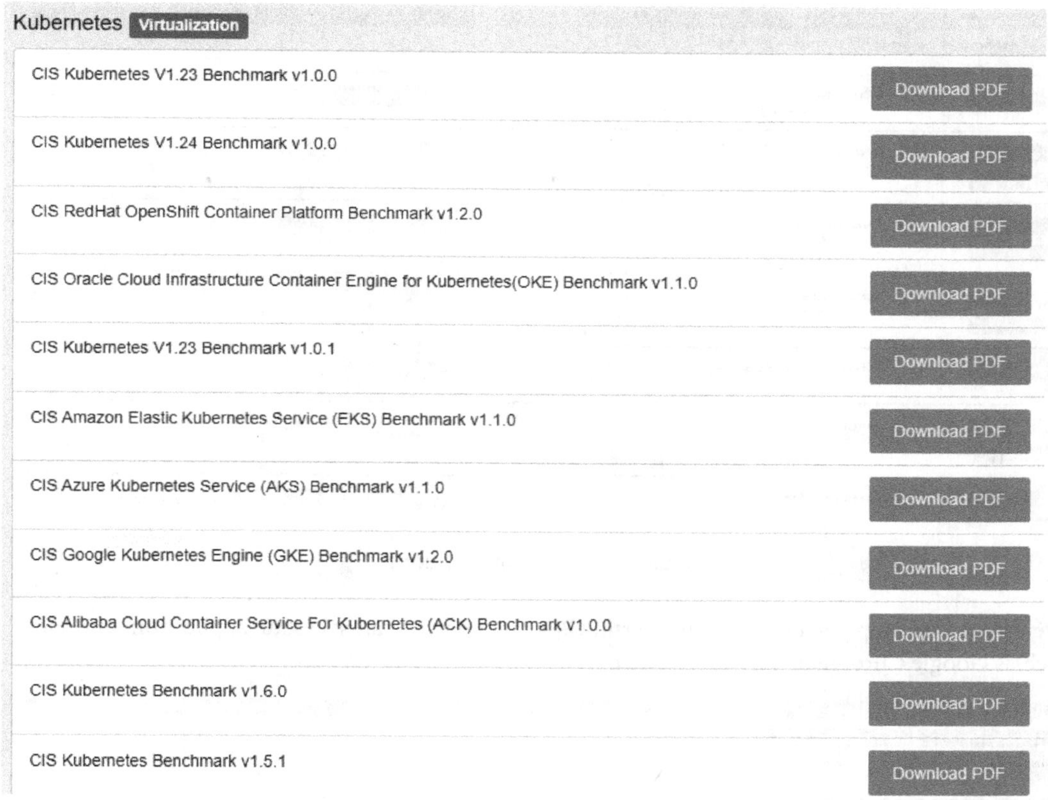

Figure 8.4 – Kubernetes CIS information

There are 302 pages, so the reality is you probably don't want to read through it all, especially after reading this chapter (or maybe you do!). Skim through it and search for things that you find interesting. I like the part about Kubernetes Secrets where it explicitly says that you should think about an external Secrets store.

A note about general server hardening

Server hardening should be an absolute priority across any environment. Whether you're running Windows servers, Linux servers, or a mixture of both, hardening your systems is the key to mitigating as much security vulnerability at the system level as possible.

Because CIS has been around for such a long time, there's a benchmark for almost everything. For example, here is a screenshot that showcases just a few benchmarks available:

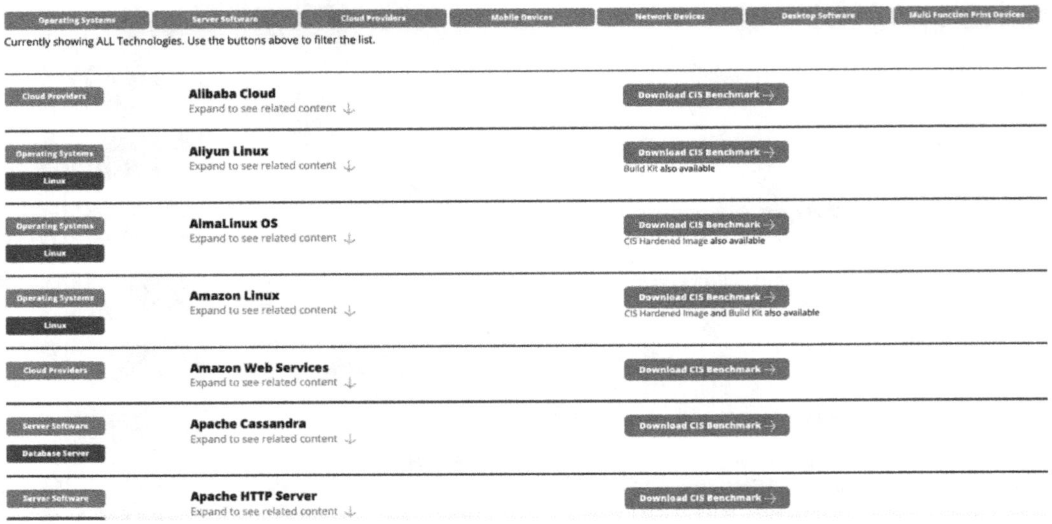

Figure 8.5 – Benchmark options

Even from a desktop perspective, you can run CIS benchmarks against certain applications and tools such as Google Chrome or Microsoft Office:

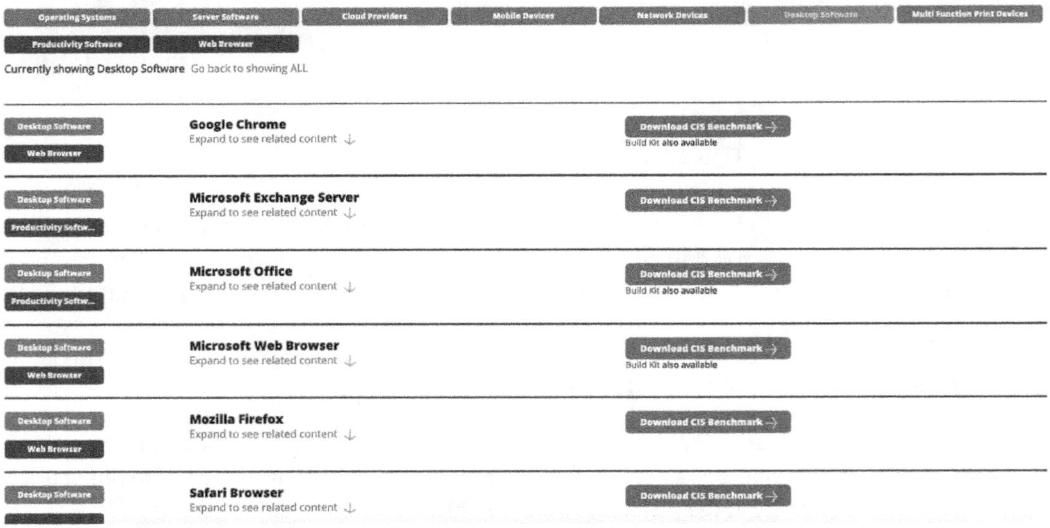

Figure 8.6 – Desktop benchmark options

To see a full list, check out https://www.cisecurity.org/cis-benchmarks/.

System scanning

Although not Kubernetes-specific, or Kubernetes-scanning-specific, the truth is that if you're running any type of system that is in your Kubernetes environment as a Control Plane, worker node, or both, you should run a system scan to ensure that the environment is properly configured. To do this, follow these steps:

1. Download the CIS-CAT® Lite tool (it's the free one) from `https://learn.cisecurity.org/cis-cat-lite`.

2. Next, extract it and open up the `Assessor-GUI` binary:

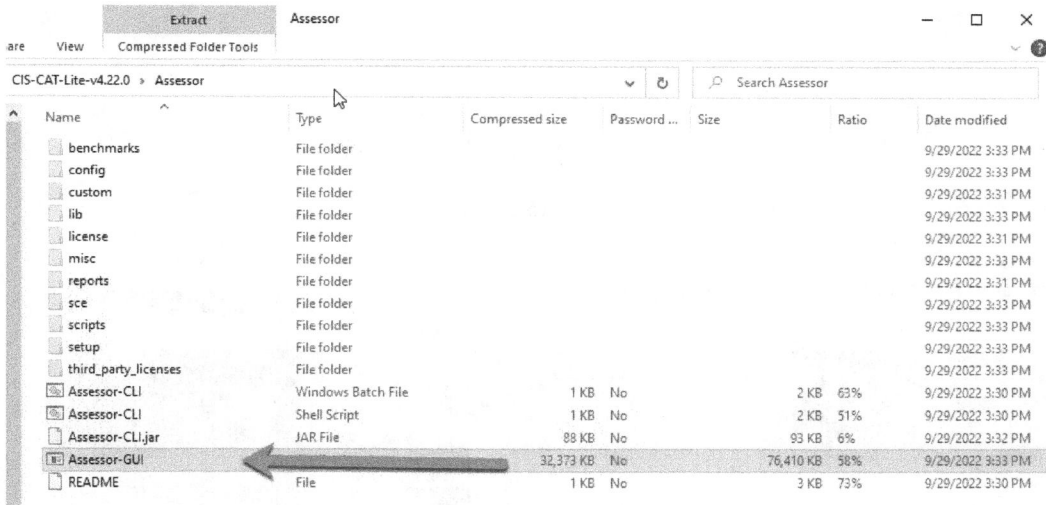

Figure 8.7 – GUI binary

3. Within the GUI tool, choose the **Advanced** option so that you can specify a remote host:

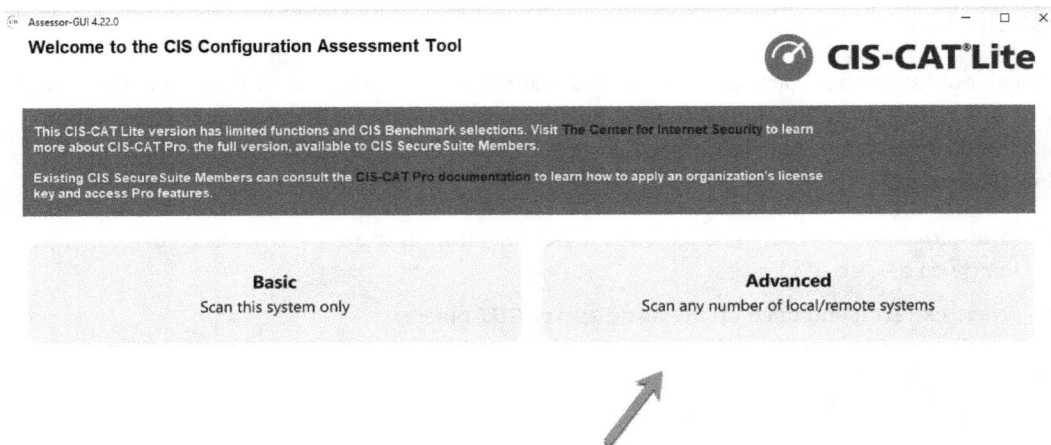

Figure 8.8 – Advanced option

4. Choose an option that gives you the ability to add a remote system:

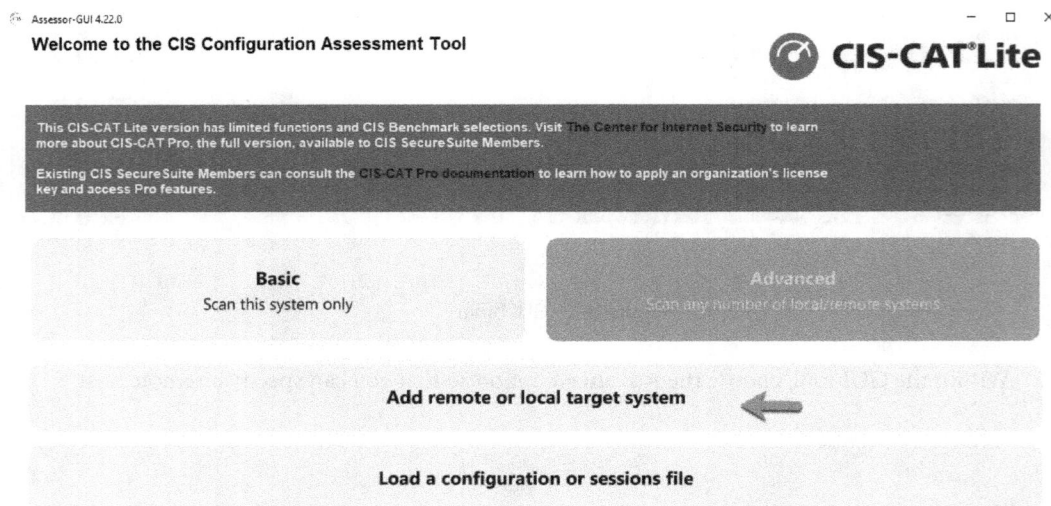

Figure 8.9 – Adding target system

5. Type in the information of the host that you wish to scan, such as the IP address, name, system type, and username/password (or SSH key):

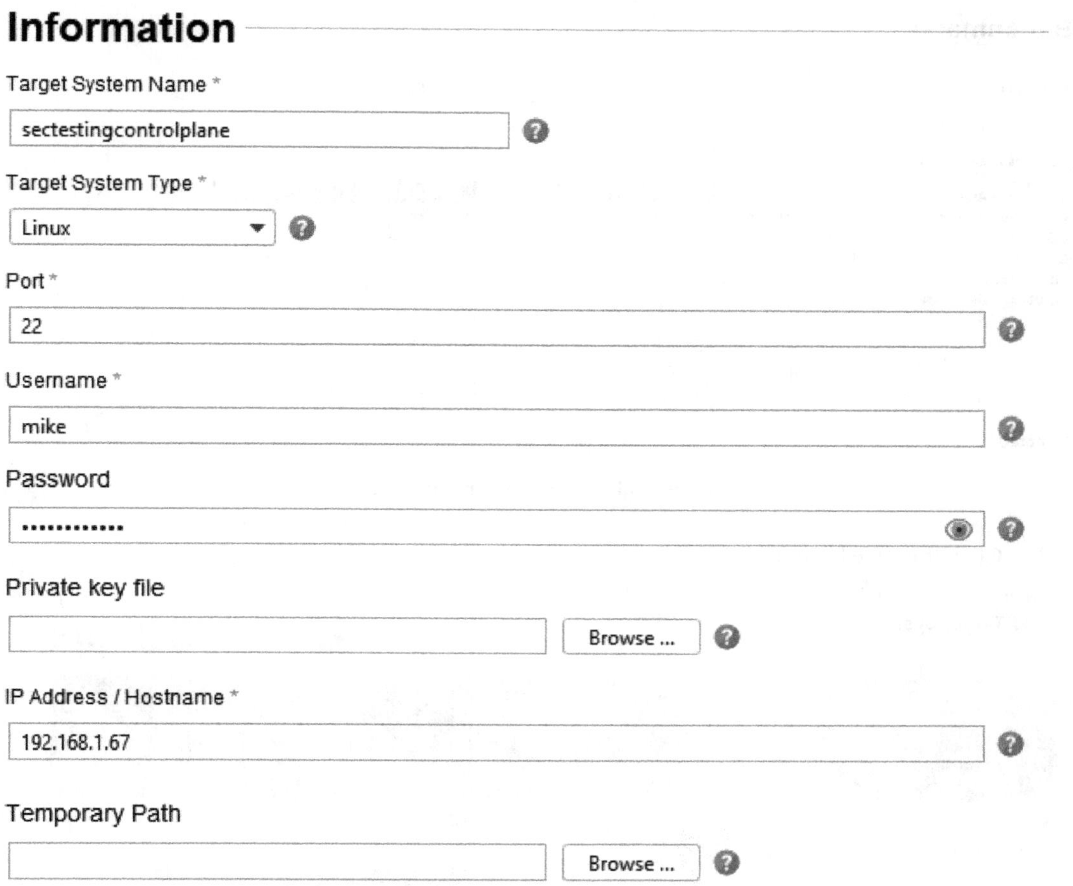

Information

Target System Name *

sectestingcontrolplane

Target System Type *

Linux ▼

Port *

22

Username *

mike

Password

••••••••••••

Private key file

Browse ...

IP Address / Hostname *

192.168.1.67

Temporary Path

Browse ...

Figure 8.10 – Target system information

6. As you can see in the next screenshot, there's no specific scan for Kubernetes. Hopefully, this will be something that's added in the future, although you'll see later in this chapter that there are tools that specifically scan Kubernetes against CIS. In this case, you can choose the Ubuntu Linux option:

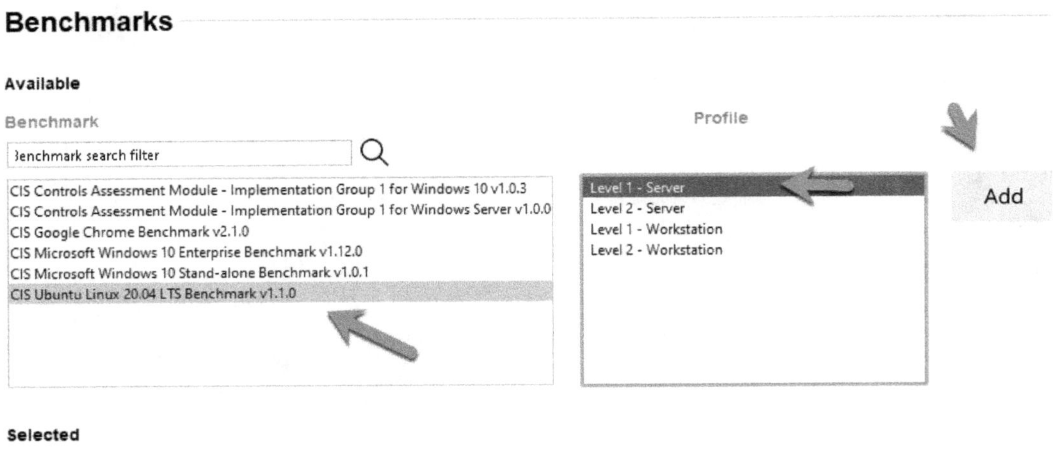

Benchmarks

Available

Benchmark

Profile

Benchmark search filter 🔍

CIS Controls Assessment Module - Implementation Group 1 for Windows 10 v1.0.3
CIS Controls Assessment Module - Implementation Group 1 for Windows Server v1.0.0
CIS Google Chrome Benchmark v2.1.0
CIS Microsoft Windows 10 Enterprise Benchmark v1.12.0
CIS Microsoft Windows 10 Stand-alone Benchmark v1.0.1
CIS Ubuntu Linux 20.04 LTS Benchmark v1.1.0

Level 1 - Server
Level 2 - Server
Level 1 - Workstation
Level 2 - Workstation

Add

Selected

Figure 8.11 – Available benchmarks

7.   Click the **Save** button:

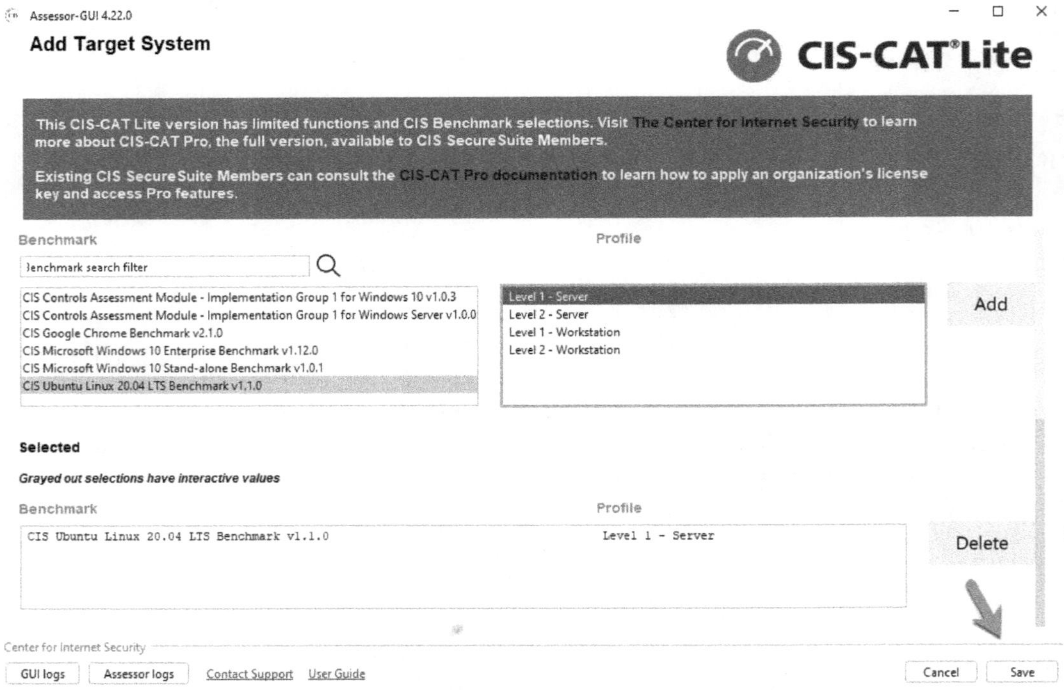

Assessor-GUI 4.22.0   —   ☐   ✕

Add Target System   ⊘ CIS-CAT®Lite

This CIS-CAT Lite version has limited functions and CIS Benchmark selections. Visit The Center for Internet Security to learn more about CIS-CAT Pro, the full version, available to CIS SecureSuite Members.

Existing CIS SecureSuite Members can consult the CIS-CAT Pro documentation to learn how to apply an organization's license key and access Pro features.

Benchmark

Profile

Benchmark search filter 🔍

CIS Controls Assessment Module - Implementation Group 1 for Windows 10 v1.0.3
CIS Controls Assessment Module - Implementation Group 1 for Windows Server v1.0.0
CIS Google Chrome Benchmark v2.1.0
CIS Microsoft Windows 10 Enterprise Benchmark v1.12.0
CIS Microsoft Windows 10 Stand-alone Benchmark v1.0.1
CIS Ubuntu Linux 20.04 LTS Benchmark v1.1.0

Level 1 - Server
Level 2 - Server
Level 1 - Workstation
Level 2 - Workstation

Add

Selected

Grayed out selections have interactive values

Benchmark

Profile

CIS Ubuntu Linux 20.04 LTS Benchmark v1.1.0

Level 1 - Server

Delete

Center for Internet Security

| GUI logs | Assessor logs | Contact Support | User Guide | | Cancel | Save |

Figure 8.12 – Adding target system

8. To ensure that you can properly scan the server, test the connection:

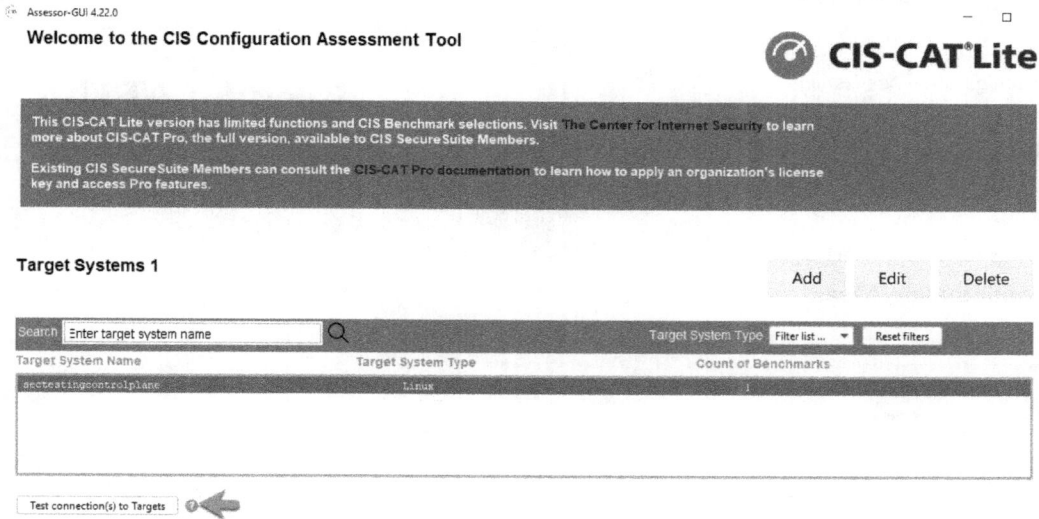

Figure 8.13 – Specifying the Control Plane

9. Click **Next**, and the testing should begin:

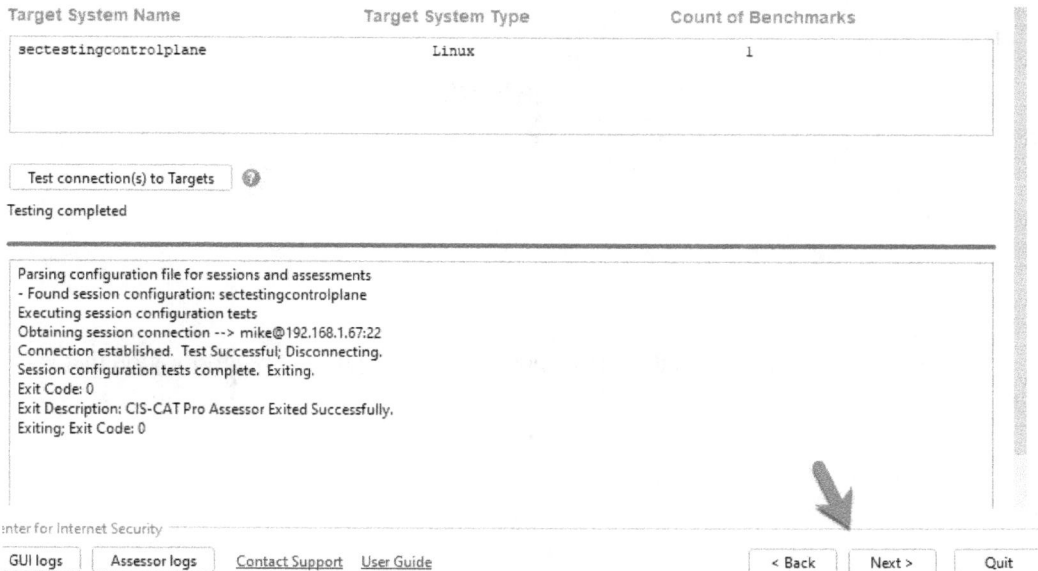

Figure 8.14 – Running the installation

10. You'll then see a screen that asks you to pick a location to save the report. Leave this at its default settings and then start the assessment:

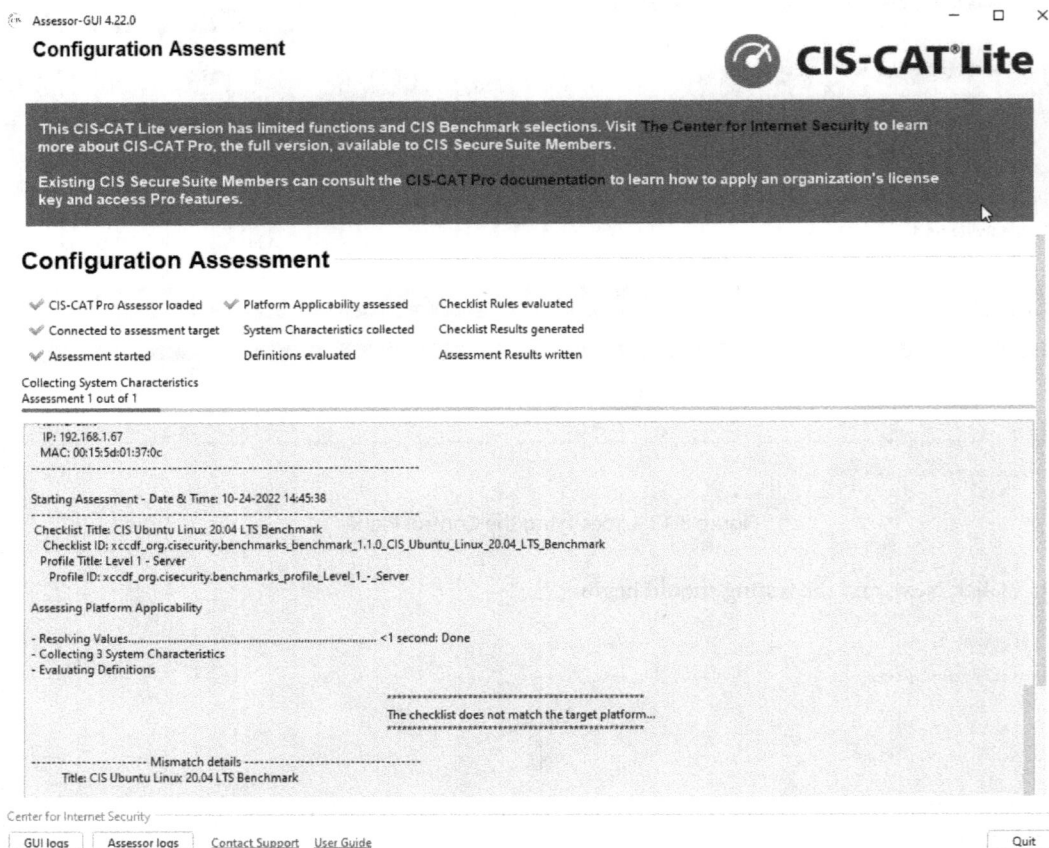

Figure 8.15 – Assessment results

Once the assessment is complete, you'll see the report output in the default report location that you saw in the prior step:

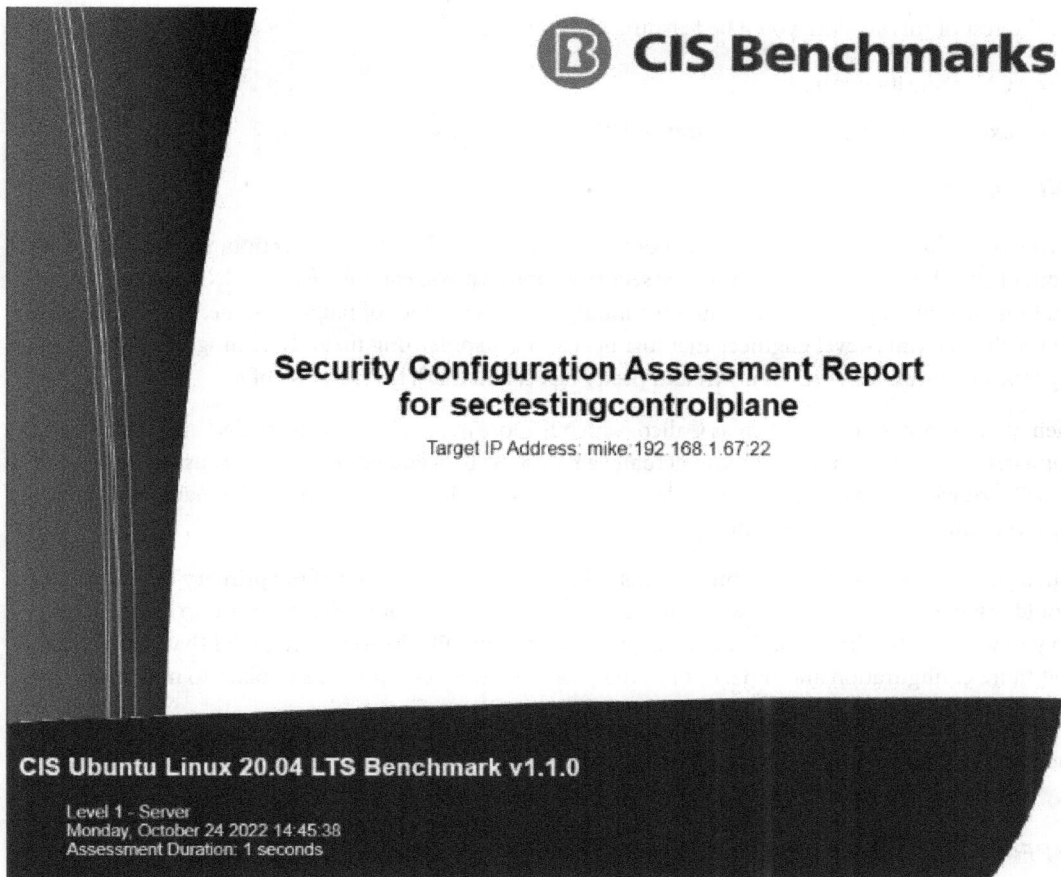

Figure 8.16 – Benchmark report

Cluster network security

In Kubernetes, there are going to be two different types of network security—internal security and host security. Host security, of course, can be anything from your cloud VPC and security groups to on-prem firewalls running in your environment. Internal security is Pod security, service security, and, overall, how Kubernetes resources communicate with each other.

To keep things Kubernetes-centric, you'll be learning about internal security and not host security. If you'd like to learn about host security, it's highly recommended to take a look at how networking works as a whole and different security-related topics such as firewalls, firewall rules, port mappings, and how network routes are configured.

For the rest of this section, you'll be learning about:

- CNI security methods

- **extended Berkeley Packet Filter (eBPF)**

CNI security

Throughout this book, you've learned about service mesh, and in the next section, you'll be learning about eBPF. There is, however, one other security approach you can take from a CNI perspective. As you look through different CNIs, you'll see multiple different types of plugins. Some, such as Flannel, are for the beginner-level engineer that just needs to get something up and running. It doesn't have any fancy features. It's watered down and pretty basic, and that's the purpose of it.

Then, you see other plugins, such as Calico, which is more of an advanced-level CNI and has a strong emphasis on security. In fact, you can actually encrypt Pod-to-Pod communication using Calico and WireGuard without even having to implement a service mesh, and that's one of the main reasons that engineers implement a service mesh.

When you're starting down your internal network security journey, one of the primary questions you should ask yourself concerns how you want to implement a CNI and why you want to implement it. Do you want a CNI that's simply *ready to go* out of the box? Or do you want a CNI that may require a bit more configuration and time, but has the proper security components in place to make your life easier in the long run?

You can learn more about Calico and WireGuard at `https://projectcalico.docs.tigera.io/security/encrypt-cluster-pod-traffic`.

eBPF

eBPF can be an entire book in itself, but in short, it's a way to remove the need to update Linux kernel code for certain programs to run. From a Kubernetes perspective, it can also remove the need for `kube-proxy`'s responsibilities.

Let's focus on a few key parts when it comes to Kubernetes and eBPF:

- Removal of `kube-proxy`

- Easier scaling

- Security

`kube-proxy` has helped make Kubernetes usable. Without it, Kubernetes wouldn't have worked. However, there's a concern. `kube-proxy` uses iptables. Although iptables have been in Linux for a long time, it doesn't scale very well. iptables rules are stored in a list, and when Pods establish a new connection to a Kubernetes Service, they go through every single iptable rule until the specific rule that's being looked for is reached. Although that may not seem like a lot for a few rules, if you have thousands (which you most likely will), it's a performance concern.

From a scalability perspective, as the number of Kubernetes Services (any type of Kubernetes Service) grows inside your cluster, the connection performance degrades. One of the reasons is that iptable rules are not incremental when you create them, which means that kube-proxy writes the whole table for every single update. It's a huge performance impact.

Now that you know some theory behind why eBPF matters, which again, can be an entire book in itself, let's dive into the hands-on implementation of eBPF:

1. First, it all depends on the cluster you're using. As with every other Kubernetes environment, if you're using a managed Kubernetes Service in the cloud, using eBPF will vary based on the CNI you specify for the Kubernetes Managed Service deployment.

   If you're planning to run Kubeadm, for example, the following command is what you should use to remove kube-proxy. Even if you don't use all the flags, ensure that you use the --skip-phases=addon/kube-proxy flag as this is needed so that kube-proxy doesn't get installed:

   ```
   sudo kubeadm init --skip-phases=addon/kube-proxy
   --control-plane-endpoint $publicIP --apiserver-advertise-
   address $ip_address --pod-network-cidr=$cidr --upload-
   certs
   ```

2. Next, install Helm if you don't already have it:

   ```
   curl -fsSL -o get_helm.sh https://raw.githubusercontent.
   com/helm/helm/main/scripts/get-helm-3
   chmod 700 get_helm.sh
   ./get_helm.sh
   ```

3. Add the Cilium Helm repo:

   ```
   helm repo add cilium https://helm.cilium.io/
   ```

4. Once the repo is added, you can install Cilium with Helm. Notice the flag to set the kube-proxy replacement:

   ```
   helm install cilium cilium/cilium \\
   --namespace kube-system \\
   --set kubeProxyReplacement=strict \\
   --set k8sServiceHost=ip_address_of_control_plane \\
   --set k8sServicePort=6443
   ```

5. After a few minutes, check to see that the Cilium Pods are running successfully by running the following command:

   ```
   kube get pods -n kube-system
   ```

The output should look similar to the following screenshot:

```
:~$ kubectl get pods -n kube-system
NAME                                READY   STATUS
cilium-8w29b                        1/1     Running
cilium-operator-67fdc9d687-8bqz6    1/1     Running
cilium-operator-67fdc9d687-s29sd    1/1     Running
cilium-r4w4g                        1/1     Running
coredns-565d847f94-cknk7            1/1     Running
coredns-565d847f94-hx8fh            1/1     Running
etcd-cpcinium                       1/1     Running
kube-apiserver-cpcinium             1/1     Running
kube-controller-manager-cpcinium    1/1     Running
kube-scheduler-cpcinium             1/1     Running
```

Figure 8.17 – Cilium Pods

Utilizing eBPF is still an extremely new topic and you may not see it through all environments. However, I can assure you that you'll begin to see it more and more as eBPF becomes more popular and the benefits are seen more.

Upgrading the Kubernetes API

In every Kubernetes environment, you must keep track of the Kubernetes API. The last thing you want to do is have an insanely out-of-date API for any software/platform, but definitely for Kubernetes as well. All APIs, even Kubernetes, can eventually have a security hole that needs to be patched. You must ensure that your environment is ready for it.

When you keep track of a Kubernetes API, the inevitable will happen: you'll have to upgrade the API. This isn't just for features and to keep the system up to date, but from a security perspective, you don't want to be too far behind as every old version of every piece of software stops getting patched and security holes open.

For the rest of this section, you'll learn how to do a Kubernetes upgrade on a cluster running Kubeadm. If you don't have Kubeadm, that's fine—still follow along. Eventually, you'll have to do an upgrade on a raw Kubernetes cluster, so it's still good to know.

> **Important note**
> For any type of upgrade, especially in production, you should not only vigorously test the upgrade path, but you should back up your environment components.

Upgrading Control Planes

Let's begin by upgrading a Kubeadm Control Plane. Follow along with these steps:

1. Run the `upgrade` command, which will show which upgrade path is available:

    ```
    kubeadm upgrade plan
    ```

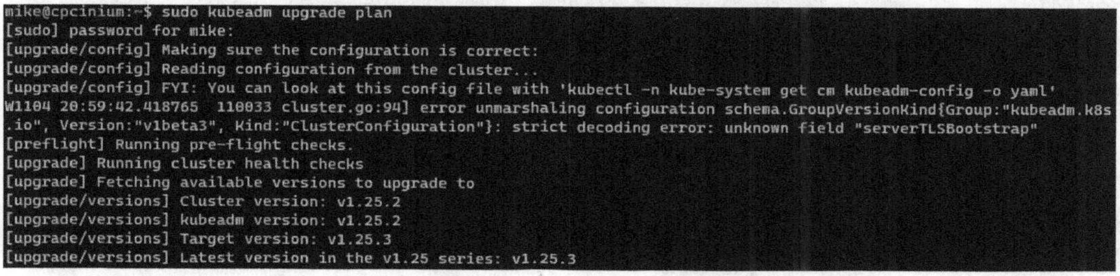

    ```
    mike@cpcinium:~$ sudo kubeadm upgrade plan
    [sudo] password for mike:
    [upgrade/config] Making sure the configuration is correct:
    [upgrade/config] Reading configuration from the cluster...
    [upgrade/config] FYI: You can look at this config file with 'kubectl -n kube-system get cm kubeadm-config -o yaml'
    W1104 20:59:42.418765  110033 cluster.go:94] error unmarshaling configuration schema.GroupVersionKind{Group:"kubeadm.k8s
    .io", Version:"v1beta3", Kind:"ClusterConfiguration"}: strict decoding error: unknown field "serverTLSBootstrap"
    [preflight] Running pre-flight checks.
    [upgrade] Running cluster health checks
    [upgrade] Fetching available versions to upgrade to
    [upgrade/versions] Cluster version: v1.25.2
    [upgrade/versions] kubeadm version: v1.25.2
    [upgrade/versions] Target version: v1.25.3
    [upgrade/versions] Latest version in the v1.25 series: v1.25.3
    ```

    Figure 8.18 – Kubernetes upgrade

2. In the following output, you'll see the target versions for every upgrade available, along with the command to run. The output will also show what the current Kubernetes API version is and which Control Plane components will be upgraded:

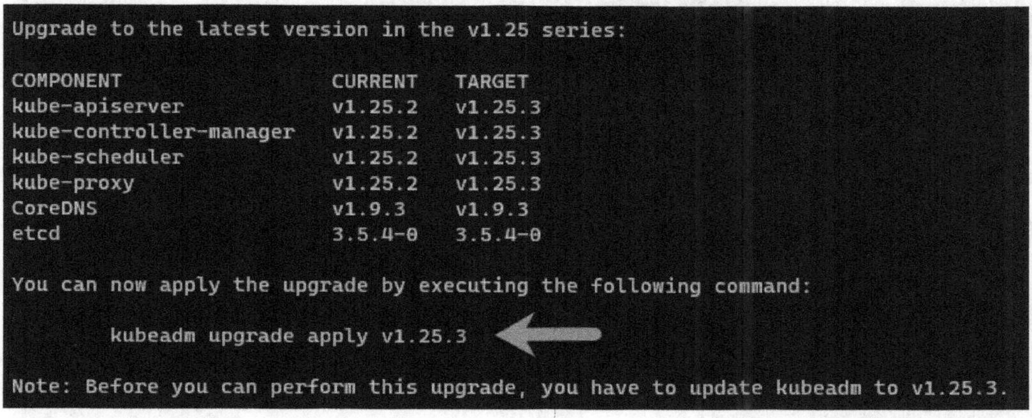

    ```
    Upgrade to the latest version in the v1.25 series:

    COMPONENT                  CURRENT    TARGET
    kube-apiserver             v1.25.2    v1.25.3
    kube-controller-manager    v1.25.2    v1.25.3
    kube-scheduler             v1.25.2    v1.25.3
    kube-proxy                 v1.25.2    v1.25.3
    CoreDNS                    v1.9.3     v1.9.3
    etcd                       3.5.4-0    3.5.4-0

    You can now apply the upgrade by executing the following command:

            kubeadm upgrade apply v1.25.3    <—

    Note: Before you can perform this upgrade, you have to update kubeadm to v1.25.3.
    ```

    Figure 8.19 – Upgrade path

3. Before running the upgrade, you'll want to download the latest version of the API and confirm that Kubeadm gets put on hold to not upgrade all Control Plane components at once. Note that running the following command may result in you having to restart the server:

    ```
    apt-mark unhold kubeadm && apt-get update && apt-get
    install -y kubeadm=1.25.x-00 && apt-mark hold kubeadm
    ```

4.   Once complete, run the upgrade, like so:

```
kubeadm upgrade apply v1.25.x
```

You'll see an output similar to the following screenshot:

Figure 8.20 – Upgrade output

Here is the second part of the output from the preceding screenshot:

Figure 8.21 – Upgrade output continued

Now that you've upgraded the Control Plane, let's learn how to upgrade worker nodes.

Upgrading worker nodes

1. Before running the upgrade, you'll want to download the latest version and confirm that Kubeadm gets put on hold to not upgrade all Control Plane components at once. Note that running the following command may result in you having to restart the server:

    ```
    apt-mark unhold kubeadm && apt-get update && apt-get
    install -y kubeadm=1.25.x-00 && apt-mark hold kubeadm
    ```

2. Next, upgrade the worker node, like so:

    ```
    sudo kubeadm upgrade node
    ```

```
mike@wpcilium:~$ sudo kubeadm upgrade node
[sudo] password for mike:
[upgrade] Reading configuration from the cluster...
[upgrade] FYI: You can look at this config file with 'kubectl -n kube-system get cm kubeadm-config -o yaml'
W1104 21:25:49.049984  110330 configset.go:78] Warning: No kubeproxy.config.k8s.io/v1alpha1 config is loaded. Continuin
g without it: configmaps "kube-proxy" is forbidden: User "system:node:wpcilium" cannot get resource "configmaps" in API
 group "" in the namespace "kube-system": no relationship found between node 'wpcilium' and this object
[preflight] Running pre-flight checks
[preflight] Skipping prepull. Not a control plane node.
[upgrade] Skipping phase. Not a control plane node.
[kubelet-start] Writing kubelet configuration to file "/var/lib/kubelet/config.yaml"
[upgrade] The configuration for this node was successfully updated!
[upgrade] Now you should go ahead and upgrade the kubelet package using your package manager.
```

Figure 8.22 – Node upgrade

That's it! This process is a bit more straightforward compared to the Control Plane.

Upgrading the kubelet

The last step is to upgrade the kubelet on both the Control Planes and worker nodes. Follow along:

1. Run the following for the kubelet upgrade:

    ```
    apt-mark unhold kubelet kubectl && apt-get update &&
    apt-get install -y kubelet=1.25.x-00 kubectl=1.25.x-00 &&
    apt-mark hold kubelet kubectl
    ```

2. Next, reload the kubelet, like so:

    ```
    sudo systemctl daemon-reload
    sudo systemctl restart kubelet
    ```

3. Run the following command, and you should now see that the Kubernetes cluster is upgraded:

    ```
    kubectl get nodes
    ```

Although this may not seem like something purely security related, and maybe it's not, it's still extremely important for security. You can't have old versions of software lying around, just as you can't have old versions of APIs lying around. For platform engineering teams, it's no different.

Audit logging and troubleshooting

Kubernetes generates several logs. In fact, most Kubernetes resources have the metrics endpoint enabled. That means, everything and anything that's generated with that Kubernetes resource—such as authentication, access, Pods going down, containers coming up, end users accessing it, and everything in between—is recorded.

The problem is that audit logging—and, sometimes, even the metrics server—isn't enabled or even installed by default. You have the ability to install and configure audit logging in Kubernetes, but it's not prepared out of the box.

What's meant by that is the Kubernetes API for audit logging is available and *turned on* out of the box. It just won't start to generate any logs that you can see because you first need to set up a policy via the `audit.k8s.io/v1` API, but policies don't exist by default—it's up to the engineer to create those policies. The policy can be anything from *show me everything* to *show me particular read actions on these particular Kubernetes resources*. It can be as high-level or as granular as you'd like.

There are a lot of policies, including audit logging, that can be turned on. In fact, it could most likely be a topic that spans an entire cluster itself. Because of that, we'll stick with audit logging in this section. However, the following screenshot showcases the **Open Web Application Security Project (OWASP)** Top 10 for Kubernetes, and one of the top 10 is proper logging:

The following logging sources should be enabled and configured appropriately:

Kubernetes Audit Logs: Audit logging is a Kubernetes feature that records actions taken by the API for later analysis. Audit logs help answer questions pertaining to events occurring on the API server itself.

Ensure logs are monitoring for anomalous or unwanted API calls, especially any authorization failures (these log entries will have a status message "Forbidden"). Authorization failures could mean that an attacker is trying to abuse stolen credentials.

Managed Kubernetes providers, including AWS, Azure, and GCP provide optional access to this data in their cloud console and may allow you to set up alerts on authorization failures.

Kubernetes Events: Kubernetes events can indicate any Kubernetes resource state changes and errors, such as exceeded resource quota or pending pods, as well as any informational messages.

Application & Container Logs: Applications running inside of Kubernetes generate useful logs from a security perspective. The easiest method for capturing these logs is to ensure the output is written to standard output `stdout` and standard error `stderr` streams. Persisting these logs can be carried out in a number of ways. It is common for operators to configure applications to write logs to a log file which is then consumed by a sidecar container to be shipped and processed centrally.

Operating System Logs: Depending on the OS running the Kubernetes nodes, additional logs may be available for processing. Logs from programs such as `systemd` are available using the `journalctl -u` command.

Cloud Provider Logs: If you are operating Kubernetes in a managed environment such as AWS EKS, Azure AKS, or GCP GKE you can find a number of additional logging streams available for consumption. One example, is within Amazon EKS there exists a log stream specifically for the Authenticator component. These logs represent the control plane component that EKS uses for RBAC authentication using AWS IAM credentials and can be a rich source of data for security operations teams.

Network Logs: Network logs can be captured within Kubernetes at a number of layers. If you are working with traditional proxy or ingress components such as nginx or apache, you should use the standard out `stdout` and standard error `stderr` pattern to capture and ship these logs for further investigation. Other projects such as eBPF aim to provide consumable network and kernel logs to greater enhance security observability within the cluster.

As outlined above, there is no shortage of logging mechanisms available within the Kubernetes ecosystem. A robust security logging architecture should not only capture relevant security events, but also be centralized in a way that is queryable, long term, and maintains integrity.

Figure 8.23 – OWASP Top 10

You can see more information about it here: `https://github.com/OWASP/www-project-kubernetes-top-ten/blob/main/2022/en/src/K05-inadequate-logging.md`.

Before jumping into the hands-on part, let's talk about what audit logging is. Audit logging is recorded by the Kubernetes API server. With those *recordings*, which are just logs, it documents a chronological set in an order that shows the sequence of actions on a Kubernetes cluster.

It generates:

- Actions taken by users
- Actions taken by Kubernetes resources
- The Control Plane itself

Essentially, audit logs allow you to ask yourself questions such as the following: 1) *What happened?* 2) *When did it happen?* 3) *How did it happen?* No question should be left unanswered as you can retrieve everything and anything about a Kubernetes cluster via the audit logs.

With that, let's learn how to set them up. Follow these steps:

1. Create a network policy such as the one shown next. For the purposes of this section, you can store it under `/etc/kubernetes/simple-policy.yaml`:

```
apiVersion: audit.k8s.io/v1
kind: Policy
rules:
- level: Metadata
```

> **Important note**
>
> If you're on a managed Kubernetes Service, such as AKS or EKS, you'll have to turn on audit logging in a different way, and it all depends on the server you're using. However, you should still read through this section as you'll end up coming across audit logs on bare-metal/VM environments (especially during this time when hybrid cloud is becoming far more popular) at some point on your Kubernetes journey.

2. Next, open up the following location via Vim or an editor of your choosing: `/etc/kubernetes/manifests/kube-apiserver.yaml`.

3. Add in the following code, as shown in *Figure 8.24*. This will give you the ability to set audit log consumption and set how long the logs are kept, which path the output of the audit logs should go to, and where your audit policy exists:

```
- --audit-log-maxage=7
- --audit-log-maxbackup=2
- --audit-log-maxsize=50


- --audit-log-path=/var/log/audit.log
- --audit-policy-file=/etc/kubernetes/simple-policy.yaml
```

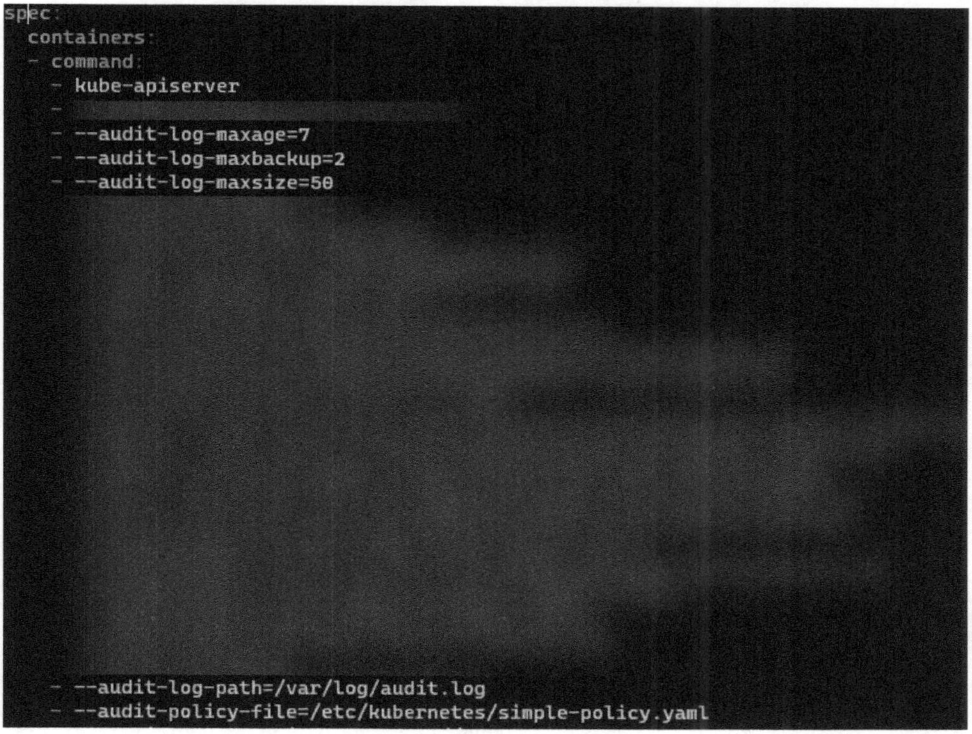

```
spec:
  containers:
  - command:
    - kube-apiserver
    -
    - --audit-log-maxage=7
    - --audit-log-maxbackup=2
    - --audit-log-maxsize=50




    - --audit-log-path=/var/log/audit.log
    - --audit-policy-file=/etc/kubernetes/simple-policy.yaml
```

Figure 8.24 – Audit policy path

4. Under `volumeMounts`, add the following code, as shown in *Figure 8.25*. For Kubernetes, the policy and the path for the audit logs need to be mounted in the cluster:

```
- mountPath: /etc/kubernetes/simple-policy.yaml
  name: audit
  readOnly: true
- mountPath: /var/log/audit.log
  name: audit-log
  readOnly: false
```

```
volumeMounts:
- mountPath: /etc/kubernetes/simple-policy.yaml
  name: audit
  readOnly: true
- mountPath: /var/log/audit.log
  name: audit-log
```

Figure 8.25 – Policy mount

5. Under hostPath, add the following:

```
- hostPath:
    path: /etc/kubernetes/simple-policy.yaml
    type: File
  name: audit
- hostPath:
    path: /var/log/audit.log
    type: FileOrCreate
  name: audit-log
```

```
volumes:
- hostPath:
    path: /etc/kubernetes/simple-policy.yaml
    type: File
  name: audit
- hostPath:
    path: /var/log/audit.log
    type: FileOrCreate
  name: audit-log
```

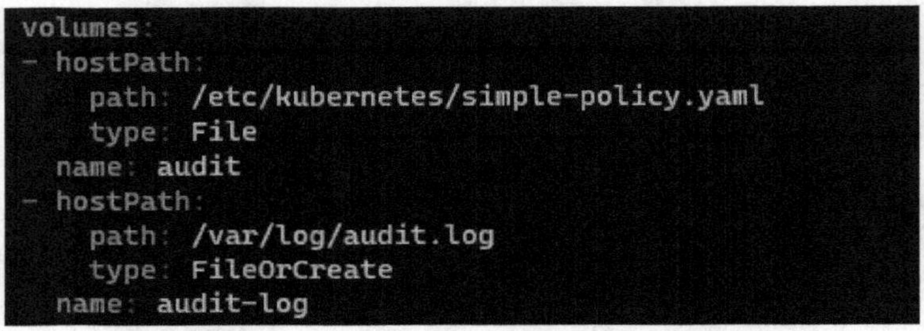

Figure 8.26 – Policy host path

6. Restart the kubelet by running the following command:

```
sudo systemctl restart kubelet
```

7. Confirm that the kubelet is still running, like so:

```
kubectl get nodes
```

You can now view the audit logs on the Control Plane at the path/location where you stored the audit.log file by executing the following command:

```
tail -f /var/log/audit.log
```

You should see a bunch of log output. For security purposes, I haven't included a screenshot showcasing the output.

As mentioned earlier, this type of configuration would be for a Kubeadm cluster or something on-prem. For the cloud, it's going to be a bit different. However, it's still important to understand this process. Remember—the cloud abstracts a lot away from engineers, but engineers must still understand the underlying components of a system to properly work with it.

Understanding RBAC

When it comes to users, groups, and service accounts, there are two questions you must ask yourself. The first is: *Who can access your cluster?* Which users, service accounts, and groups have the ability to run `kubectl` commands on the clusters in development, staging, and production? Which of those users have a Kubeconfig that gives them access to particular clusters? Which environments can they connect to?

The second question is: *What can they do once they're inside the cluster?* Can they list Pods? Create Pods? See Ingress Controllers? Create Ingress Controllers? What types of Kubernetes resources can they interact with throughout each environment?

When you're setting up a Kubernetes environment, you must also think about authentication and authorization. Who can access your cluster and what can they do? Further, you must think about what the users can do throughout each environment. For example, thinking about the single tenancy model that you learned about in a previous chapter, one engineer may have full admin access on one cluster and read-only access on another. With that, you must also think about multiple authorization methods in terms of which permissions you're giving people.

In this section, you're going to learn how to manage from a permissions perspective users, groups, and teams in Kubernetes using RBAC.

Please note that although this section is not huge, it should point you in the right direction in terms of how to think about RBAC and how to start implementing it.

What is RBAC?

RBAC, as with many other topics in this book (I'm a broken record at this point), can be an entire book in itself. Because of that, let's do a brief theoretical explanation and then dive into the hands-on piece.

RBAC, by definition, is a way to ensure that users, groups, and service accounts only have the permissions that they need from an authorization perspective. RBAC does not do authentication—it does authorization. The authentication piece comes before RBAC. Once there's a user, group, or service account created, then RBAC can jump into action and start creating permissions.

Within RBAC for Kubernetes, you have four primary resources that you want to utilize:

- `Roles`
- `ClusterRoles`
- `RoleBindings`
- `ClusterRoleBindings`

You'll learn more about them in the upcoming sections.

When you're thinking about RBAC, think: *What am I allowing this person to do inside Kubernetes?*

One thing to keep in mind is that RBAC is typically the bane of every security engineer's existence. It's one of those topics in Kubernetes that makes everyone bang their head against a wall because it can start to become insanely complex, and there's no central way to manage hundreds of RBAC roles and permissions. There are tools and platforms out there that are trying to mitigate this, such as Kubescape's RBAC Visualizer.

To continue along with this chapter, you'll need a user, group, or service account. Because Kubernetes doesn't have an out-of-the-box method for creating users and groups, let's use a service account.

Run the following command to create a new service account called `miketest`:

```
kubectl create sa miketest
```

Once the service account is created, it can be used for the following sections.

Roles and ClusterRoles

`Roles` are permissions that you can give users, groups, and service accounts, and they are namespace scoped. Meaning, let's say you create a role called `readpods`. That role would be tied to a namespace—for example, a namespace called `ingress`. That means the `readpods` role only works on the `ingress` namespace, and it's not tied to any other namespace.

How about if you want a role/permissions for a user/group/service account that's used across all namespaces through the cluster? That's where `ClusterRoles` come into play. A `ClusterRole` is the same thing as a `Role`. The only difference is that it's not namespace scoped.

Let's dive in to learn how you can create `Roles` and `ClusterRoles`.

Roles

The following code snippet is an example of a `Role` that you can create. It's scoped to the `ingress` namespace and sets read-only permissions for the Pod Kubernetes resource. Notice in the verbs that it's all read permissions—`get`, `watch`, and `list`:

```
kind: Role
apiVersion: rbac.authorization.k8s.io/v1
metadata:
  namespace: ingress
  name: reader
rules:
```

```
- apiGroups: [""]
  resources: ["pods"]
  verbs: ["get", "watch", "list"]
```

Implementing the preceding `Role` will ensure that you have a proper role created to give a user/group/service account read-only permissions for Pods in the `ingress` namespace.

ClusterRoles

As with the preceding `Role`, the following `ClusterRole` creates a `ClusterRole` called `reader` for read-only permissions on Pods. The key difference is that it's not scoped to a particular namespace:

```
kind: ClusterRole
apiVersion: rbac.authorization.k8s.io/v1
metadata:
  name: reader-cluster
rules:
- apiGroups: [""]
  resources: ["pods"]
  verbs: ["get", "watch", "list"]
```

Implementing the preceding `ClusterRole` will ensure that you have a proper role created to give a user/group/service account read-only permissions for Pods across all namespaces in the cluster.

Next, let's learn how to bind `Roles` to a particular service account.

RoleBindings and ClusterRoleBindings

A `RoleBinding` is a way that you tie/attach a `Role` to a user/group/service account. For example, let's say you have a `Role` called `podreaders` and you want to tie/attach that role to the `miketest` service account. You would use a `RoleBinding` to perform that action.

Just as with `Roles` and `ClusterRoles`, the only difference is that `RoleBindings` are namespace scoped and `ClusterRoleBindings` are not and can be used throughout the cluster.

Let's learn how to implement `RoleBindings` and `ClusterRoleBindings`.

RoleBinding

The following `RoleBinding` takes the `Role` that you created in the previous section and attaches it to the `miketest` service account. See how there's a `kind` and the service account kind is specified? This is where you can specify a `group` or a `user`. It's also scoped to the `ingress` namespace as this is not a `ClusterRoleBinding`:

```
apiVersion: rbac.authorization.k8s.io/v1
kind: RoleBinding
metadata:
  name: reader-pod
  namespace: ingress
subjects:
- kind: ServiceAccount
  name: miketest
  apiGroup: ""
roleRef:
  kind: Role
  name: reader
  apiGroup: rbac.authorization.k8s.io
```

ClusterRoleBinding

Much as with the preceding `RoleBinding`, the following `ClusterRoleBinding` will attach the `miketest` service account to the `ClusterRole` and reference the following `ClusterRole`:

```
apiVersion: rbac.authorization.k8s.io/v1
kind: ClusterRoleBinding
metadata:
  name: read-pod-global
subjects:
- kind: ServiceAccount
  name: miketest
  apiGroup: ""
roleRef:
  kind: ClusterRole
  name: reader-cluster
  apiGroup: rbac.authorization.k8s.io
```

Now that you know about overall authentication and authorization permissions, it's time to learn about overall Kubernetes resource security and the approaches that you can take out of the gate to ensure a successful security-centric deployment.

Kubernetes resource (object) security

Throughout this chapter, you learned a little bit about Kubernetes resource security. Remember, Kubernetes resources can be anything from Pods to Ingress Controllers to Services. Essentially, anything running inside of the Kubernetes cluster that you're reaching via the API is a Kubernetes resource.

In this section, you're going to learn the top methods of today to secure Kubernetes resources within Kubernetes and by using third-party tools.

Pod security

When it comes to network security in a Kubernetes environment, there are two parts—the host network and the internal network. For the purposes of this section, we can't go into host networking because every environment is going to be different. Whether it's different physical hardware or virtual hardware setups, there's no *one-size-fits-all* network environment.

However, there are a few helpful tips that work across every environment:

1. Ensure that you have proper firewall rules.
2. Ensure that you're implementing proper routing protocols and not just opening up the entire network.
3. Ensure that you have the proper port setup in place.
4. Ensure that you're logging and observing network traffic.

For Kubernetes network security, there are network policies.

Network Policies are built into Kubernetes via the `networking.k8s.io/v1` API. Network Policies act like firewall rules for both Ingress and Egress traffic. However, network policies aren't just about whitelisting or blacklisting IP addresses and ports. You can do much more with a policy. For example, you can block traffic from a specific network to a specific namespace, from a specific namespace, or to/from a specific application. Because of the vast number of options that come with Network Policies, you have plenty of options, but you'll want to ensure that you're setting up the right policies. One wrong accidental `162.x.x.x` instead of `172.x.x.x` can completely throw off the entire network workflow in a network policy and completely halt application workloads.

Let's dive into what a network policy looks like.

To test this out, run the following Pods in your Kubernetes environment:

```
kubectl run busybox1 --image=busybox --labels app=busybox1 --
sleep 3600
kubectl run busybox2 --image=busybox --labels app=busybox2 --
sleep 3600
```

The preceding new Pods will run a container image called `busybox`, which is a small form factor that's usually used for testing.

Next, obtain the IP address of the Pods, like so:

```
kubectl get pods -o wide
```

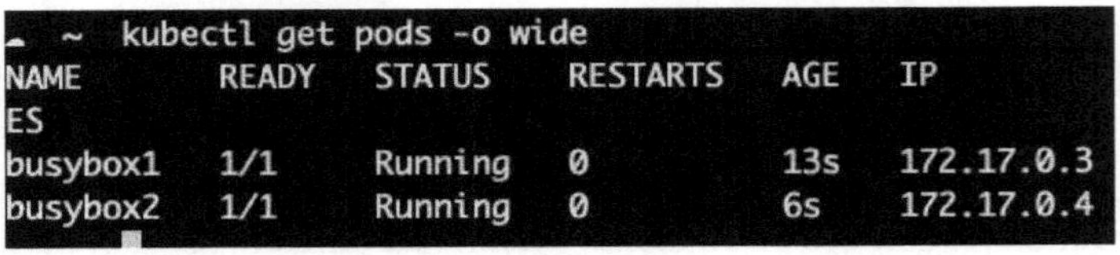

```
 ~   kubectl get pods -o wide
NAME        READY    STATUS     RESTARTS    AGE    IP
ES
busybox1    1/1      Running    0           13s    172.17.0.3
busybox2    1/1      Running    0           6s     172.17.0.4
```

Figure 8.27 – Pod output

Run a ping against the `busybox1` Pod:

```
kubectl exec -ti busybox2 -- ping -c3 ip_of_busybox_one
```

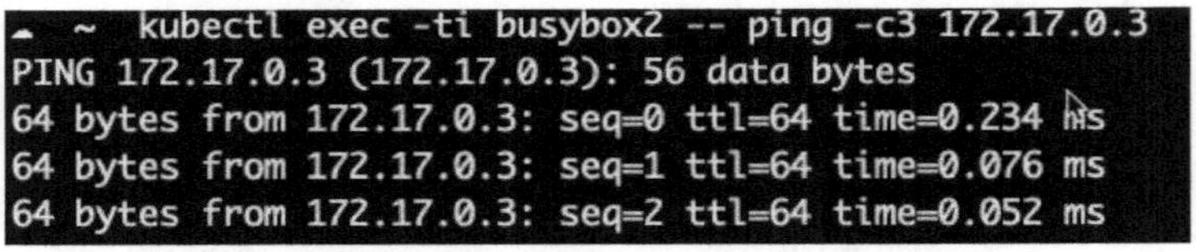

```
 ~   kubectl exec -ti busybox2 -- ping -c3 172.17.0.3
PING 172.17.0.3 (172.17.0.3): 56 data bytes
64 bytes from 172.17.0.3: seq=0 ttl=64 time=0.234 ms
64 bytes from 172.17.0.3: seq=1 ttl=64 time=0.076 ms
64 bytes from 172.17.0.3: seq=2 ttl=64 time=0.052 ms
```

Figure 8.28 – Ping output

Now that you know there's proper **Internet Control Message Protocol** (**ICMP**) communication, create a network policy that blocks all Ingress traffic to the `busybox1` Pod:

```
kubectl create -f - <<EOF
kind: NetworkPolicy
apiVersion: networking.k8s.io/v1
metadata:
  name: web-deny-all
spec:
  podSelector:
```

```
    matchLabels:
        app: busybox1
    ingress: []
EOF
```

Run the ping against `busybox1` again:

```
kubectl exec -ti busybox2 -- ping -c3 ip_of_busybox_one
```

There should now be 100% packet loss.

If you're in an environment where this didn't work, such as a standard Minikube environment, the reason why it's most likely not working is that you're using a CNI that doesn't have network policies enabled or doesn't support network policies.

To find out how to enable network policies, you'll need to do a quick search on how to implement network policies for your specific CNI.

Policy enforcement

In the previous section, you learned about security at the network layer, which is of course needed. After (or before) the network layer is the application layer, which is where policy enforcement around Pods and containers comes into play.

The whole idea behind policy enforcement is to give you the ability to protect your Pods, ensure best security practices, and set standards for your organization.

For example, one of the biggest best practices in production is to ensure that you're not using the latest container image version in production. Instead, you always want to use a container image version that's properly versioned and battle-tested for protection. With policy enforcement in Kubernetes, you can accomplish that.

Right now, the two biggest ways to implement policy enforcement are using **Open Policy Agent (OPA)** and Kyverno. They're both the same from a policy enforcement perspective, but the biggest difference is that Kyverno only works inside of Kubernetes. Because of that, a lot of engineers are going toward using OPA so that they can use it throughout their environment and not just in Kubernetes.

Because of that, the hands-on section will be using OPA.

> **What about Pod Security Policies?**
>
> If you've heard of Pod Security Policies, they're essentially the same thing as OPA. However, they were deprecated in v1.21 of Kubernetes and completely removed in v1.25.

OPA

When you want to configure a specific policy, you can use a policy agent such as OPA. OPA allows you to write policies in an OPA-specific language called Rego (which you'll see later in this section). When you write a policy, any request or event that comes in from another Kubernetes resource or an outside entity will be queried. OPA's decision agent will give it a *pass* or *fail*.

But how does OPA know how to implement policies?

That's where OPA Gatekeeper comes into play. Gatekeeper is a *middle ground* of sorts that allows Kubernetes to interact with OPA. Gatekeeper is installed on Kubernetes and enables the use of OPA policies.

Let's dive in from a hands-on perspective to set up OPA. The first part will be deploying OPA Gatekeeper and the second part will be implementing a policy. Proceed as follows:

1. Add Helm chart for Gatekeeper, like so:

    ```
    helm repo add gatekeeper https://open-policy-agent.
    github.io/gatekeeper/charts
    ```

2. Install the Helm chart by running the following command:

    ```
    helm install gatekeeper/gatekeeper --name-
    template=gatekeeper --namespace gatekeeper-system
    --create-namespace
    ```

3. Confirm that all Kubernetes resources for Gatekeeper were deployed:

    ```
    kubectl get all -n gatekeeper-system
    ```

Now that Gatekeeper is installed, let's start implementing the OPA policies.

The configuration is the definition/output of what OPA Gatekeeper is allowed to create policies for. In the following `config.yaml` file, because of the way that it's written, Gatekeeper knows that it can only specify policies for Pods and no other Kubernetes resources. Run the following code:

```
kubectl create -f - <<EOF
apiVersion: config.gatekeeper.sh/v1alpha1
kind: Config
metadata:
  name: config
  namespace: "gatekeeper-system"
spec:
  sync:
```

```
    syncOnly:
      - group: ""
        version: "v1"
        kind: "Pod"
EOF
```

A constraint template is a policy that you configure for an environment. It's a template, so you can use it across multiple places.

The Rego code/policy in the following constraint template ensures no one can utilize the latest tag of a container image. Run the following code:

```
kubectl create -f - <<EOF
apiVersion: templates.gatekeeper.sh/v1beta1
kind: ConstraintTemplate
metadata:
  name: blocklatesttag
  annotations:
    description: Blocks container images from using the latest
tag
spec:
  crd:
    spec:
      names:
        kind: blocklatesttag # this must be the same name as
the name on metadata.name (line 4)
  targets:
    - target: admission.k8s.gatekeeper.sh
      rego: |
        package blocklatesttag
        violation[{"msg": msg, "details": {}}]{
        input.review.object.kind == "Pod"
        imagename := input.review.object.spec.containers[_].
image
        endswith(imagename,"latest")
        msg := "Images with tag the tag \"latest\" is not
allowed"
        }
EOF
```

Next, you have the constraint. The constraint takes the template that you created earlier and allows you to use the template to create a policy inside of a Kubernetes cluster:

```
kubectl create -f - <<EOF
apiVersion: constraints.gatekeeper.sh/v1beta1
kind: blocklatesttag
metadata:
  name: nolatestcontainerimage
spec:
  match:
    kinds:
      - apiGroups: [""]
        kinds: ["Pod"]
  parameters:
    annotation: "no-latest-tag-used"
EOF
```

The OPA policy is now created.

To confirm that the policy is working as expected, you can test it out with the two following Kubernetes manifests:

- The following manifest with the container image's latest tag shouldn't work because of the policy that you created earlier. The deployment itself will deploy, but the Pods won't be scheduled and won't come online.

  Try running the following code:

  ```
  kubectl create -f - <<EOF
  apiVersion: apps/v1
  kind: Deployment
  metadata:
    name: nginx-deployment
  spec:
    selector:
      matchLabels:
        app: nginxdeployment
    replicas: 2
    template:
      metadata:
  ```

```
        labels:
          app: nginxdeployment
      spec:
        containers:
        - name: nginxdeployment
          image: nginx:latest
          ports:
          - containerPort: 80
  EOF
```

Wait a few minutes and when you see that it doesn't come online, delete it, as follows:

```
kubectl delete deployment nginx-deployment
```

- Next, try the following manifest. It will work, and the Pods will come online because the container image version is specified:

```
kubectl create -f - <<EOF
apiVersion: apps/v1
kind: Deployment
metadata:
  name: nginx-deployment
spec:
  selector:
    matchLabels:
      app: nginxdeployment
  replicas: 2
  template:
    metadata:
      labels:
        app: nginxdeployment
    spec:
      containers:
      - name: nginxdeployment
        image: nginx:1.23.1
        ports:
        - containerPort: 80
  EOF
```

OPA is a huge topic in itself. I highly recommend diving into it more. We only had a few pages together in this book to dive into it, but it goes far more in-depth.

Scanning container images

One popular security-style *entry point* for many engineers to start their security journey is by scanning container images. Scanning a container image means that you're using a tool/platform to look inside the container image and see if there are any vulnerabilities. The vulnerability list typically comes from the NVD and the CIS benchmarks for Kubernetes. Both are a curated list of best practices from a security perspective and also contain known vulnerabilities.

There are a lot of tools in this space. In this section, let's stick to one that's as *built in* as possible: Snyk.

Snyk is used to scan containers for vulnerabilities from a list that's pre-defined (as stated earlier) of best practices. A while back, Docker and Snyk partnered to ensure that security is embedded natively into any containerized workload. With that partnership, when you run the `docker scan` command, it's actually using Snyk on the backend.

To use Snyk, ensure that you have the Docker CLI installed and run the following command:

```
docker scan containerimage:containerversion
```

For example, let's say you want to scan the `ubuntu:latest` container image, as shown here:

```
docker scan ubuntu:latest
```

Once you run the `docker scan` command, you can scroll through all of the vulnerabilities. You'll see a summary of the vulnerabilities that were found, what was tested, and which platform was used.

Vulnerabilities can range from being super basic, in that it just ends up being a best practice to fix, or something that's incredibly crucial and leaves your environment open for attack.

Kubernetes Secrets

Wrapping up this chapter, and the overall book, you'll learn about Kubernetes Secrets.

Secrets, in short, are anything that you don't want to be in plain text. Typically, they are things such as passwords and API keys. However, they could even be usernames. Any type of data that you don't want to be in plain text, at rest, or in transit can be considered a Secret.

At this point in your engineering journey, it's assumed that you don't need to be taught about Secrets, so we're going to skip that part and dive right into the hands-on part.

Creating Kubernetes Secrets

To create a Kubernetes Secret, you'll use the `secret` resource from the `v1` core API group.

For example, the following is a Secret called `testsecret` with a username and password:

```
apiVersion: v1
kind: Secret
metadata:
  name: testsecret
type: Opaque
data:
  username: YWRtaW4=
  password: MWYyZDFlMmU2N2Rm
```

Confirm that the Secret was created by running the following command:

```
kubectl get secrets
```

Next, use the `secret` by putting it inside a Pod, like so:

```
apiVersion: v1
kind: Pod
metadata:
  name: nginxpod
spec:
  containers:
  - name: mypod
    image: nginx:latest
  volumes:
  - name: foo
    secret:
      secretName: testsecret
```

Don't use Kubernetes Secrets

Although you literally just created a new Kubernetes Secrets a few seconds ago, here's the thing—it's not a recommended practice.

The biggest reason is that the default opaque standard for Kubernetes Secrets stores secrets in plain text. Yes—that's right. The secrets will be stored in plain text in the `etcd` database. Thinking about it from another perspective, think about Kubernetes Manifests. Even if the secret wasn't in plain text in Etcd, it would still be in plain text in the Kubernetes Manifest that's creating the secret, and if it's in plain text, where would you store it? You can't push the manifest up to GitHub because then your secret would be compromised. Because of this, many engineers—and, quite frankly, even the Kubernetes documentation—highly recommend using a third-party secret provider. The most popular at this time for Kubernetes is HashiCorp Vault.

Summary

As you went through this chapter, there may have been some thoughts in your head of pure confusion. That's okay—we're all trying to *get it* when it comes to security in general, especially in Kubernetes.

Kubernetes security is an advanced topic, which is why the goal was to leave this topic for the last chapter of the book. Without Kubernetes security, environments will continue to be targets for attackers. However, before understanding Kubernetes security, you must fully understand how to utilize Kubernetes in production. The goal of *Chapters 1-7* was to help with understanding Kubernetes in production.

The next goal, once you close this book, is to take what you've learned in this chapter along with the various methodologies highlighted and implement them in your production environment for optimal results.

Further reading

- *Learn Kubernetes Security* by *Kaizhe Huang* and *Pranjal Jumde*: `https://www.packtpub.com/product/learn-kubernetes-security/9781839216503`

Index

A

C

‹packt›

Other Books You May Enjoy

If you enjoyed this book, you may be interested in these other books by Packt:

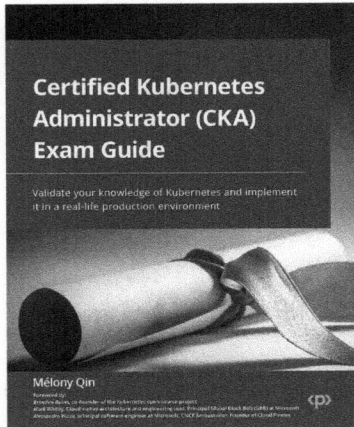

Certified Kubernetes Administrator (CKA) Exam Guide

Mélony Qin

ISBN: 978-1-80323-826-5

- Understand the fundamentals of Kubernetes and its tools
- Get hands-on experience in installing and configuring Kubernetes clusters
- Manage Kubernetes clusters and deployed workloads with ease
- Get up and running with Kubernetes networking and storage
- Manage the security of applications deployed on Kubernetes
- Find out how to monitor, log, and troubleshoot Kubernetes clusters and apps among others

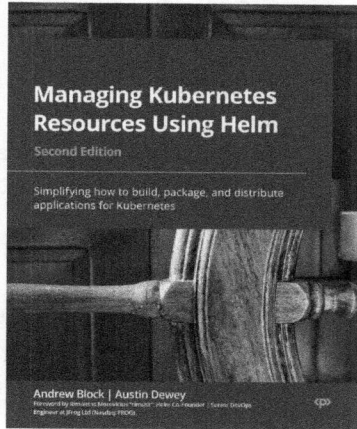

Managing Kubernetes Resources Using Helm - Second Edition

Andrew Block, Austin Dewey

ISBN: 978-1-80324-289-7

- Understand how to deploy applications on Kubernetes with ease
- Package dynamic applications for deployment on Kubernetes
- Integrate Helm into an existing software release process
- Develop an enterprise automation strategy on Kubernetes using Helm
- Use Helm within a Helm Kubernetes operator
- Leverage Helm in a secure and stable manner that fits the enterprise
- Discover the ins and outs of automation with Helm

Packt is searching for authors like you

If you're interested in becoming an author for Packt, please visit `authors.packtpub.com` and apply today. We have worked with thousands of developers and tech professionals, just like you, to help them share their insight with the global tech community. You can make a general application, apply for a specific hot topic that we are recruiting an author for, or submit your own idea.

Share Your Thoughts

Now you've finished *50 Kubernetes Concepts Every DevOps Engineer Should Know*, we'd love to hear your thoughts! Scan the QR code below to go straight to the Amazon review page for this book and share your feedback or leave a review on the site that you purchased it from.

`https://packt.link/r/1804611476`

Your review is important to us and the tech community and will help us make sure we're delivering excellent quality content.

Download a free PDF copy of this book

Thanks for purchasing this book!

Do you like to read on the go but are unable to carry your print books everywhere?

Is your eBook purchase not compatible with the device of your choice?

Don't worry, now with every Packt book you get a DRM-free PDF version of that book at no cost.

Read anywhere, any place, on any device. Search, copy, and paste code from your favorite technical books directly into your application.

The perks don't stop there, you can get exclusive access to discounts, newsletters, and great free content in your inbox daily

Follow these simple steps to get the benefits:

1. Scan the QR code or visit the link below

https://packt.link/free-ebook/9781804611470

2. Submit your proof of purchase
3. That's it! We'll send your free PDF and other benefits to your email directly

www.ingramcontent.com/pod-product-compliance
Lightning Source LLC
Chambersburg PA
CBHW061351210326
41598CB00035B/5952